Antenna Transfer Function
&
LFM RADAR
Wide Band Performance Analysis of Wire Antenna & Design of Matched Filter for LFM RADAR

Mainak Mukhopadhyay

Chapter 1

Wide Band Performance Analysis of Wire Antenna in Transmitting and Receiving Mode

This chapter deals with the evaluation of the phase versus frequency response of a wide-band electromagnetic signal transmitted / received by a wire antenna. A system with a transmitting and receiving antenna is basically a two-port network and can be represented by the S-parameters. In a complete communication system when a signal is transmitted by a transmitting antenna and received by a receiving antenna, the complex S_{21} parameter of the two-port network gives the complete frequency response of the system. The phase versus frequency characteristics plays an important role and knowledge of the same is extremely essential for dealing with practical waveforms, which are non-sinusoidal and hence wide band in nature.

1.1 Introduction

Wire antennas are used for the time-domain measurements of transient electromagnetic fields caused by electrostatic discharges. In this case, the output signal waveform of a sensor may be different from the input electromagnetic waveform due to the frequency characteristics of the employed antenna and circuits. The output waveform is represented as a convolution integral of the input waveform and the impulse response of the antenna system. If the impulse response or its Fourier Transform (Transfer Function) can be determined, the input waveform is reconstructed by the deconvolution technique. In transient field measurements, an antenna with wideband performance both in amplitude and in phase is desired. To characterize such an antenna, Ishigami, Iwasaki, defined the term Complex Antenna Factor (CAF). In analogy to the traditional scalar antenna factor, it relates the output voltage of the antenna to the incident electric field. The CAF adds phase values to the conventional scalar antenna factor. The CAF for an incident plane wave of angular frequency ω (with the polarization giving maximum output) is the ratio of the incident complex electric field strength at the point of an antenna element to the complex matched output voltage of the antenna for the load of the matched impedance. They had measured and computed CAFs for linear monopoles by near field 3-antenna method involving complicated measurements . H. Hosoyama, Iwasaki,evaluated the CAF of a V-dipole antenna with a balun consisting of two coaxial feeders. The effective length and input impedance were calculated by using the Galerkin's moment method along with the delta-gap driving model. To obtain the effective length in the receiving mode, the open–circuit voltage at the driving point was calculated from an incident plane wave electric field on the V-dipole antenna.

The CAF allows the determination of both the amplitude and phase information of the incident field and also for finding the direction of arrival of electromagnetic waves . Knowledge of the Complex Antenna Factor of an antenna over a broad frequency range allows the reconstruction of complicated, non-sinusoidal time domain waveforms from the predicted data . Such fields are typical for the radiated emissions of a complex, digital system such as a computer or a mixed signal system. In principle, any physical band pass waveform may be recovered from measurements made with an antenna if the spectrum of the signal falls within the operating frequency range of the antenna and the CAF is known over this frequency range.

In this chapter, a Method of Moment based numerical technique has been applied to evaluate the current distribution and hence the phase variation with frequency of the transmitted and received signal. The antenna elements chosen are dipole antennas and loaded antennas (e.g. inverted L, T, I and C-shaped antennas). Also the overall response of the two-port network has been studied.

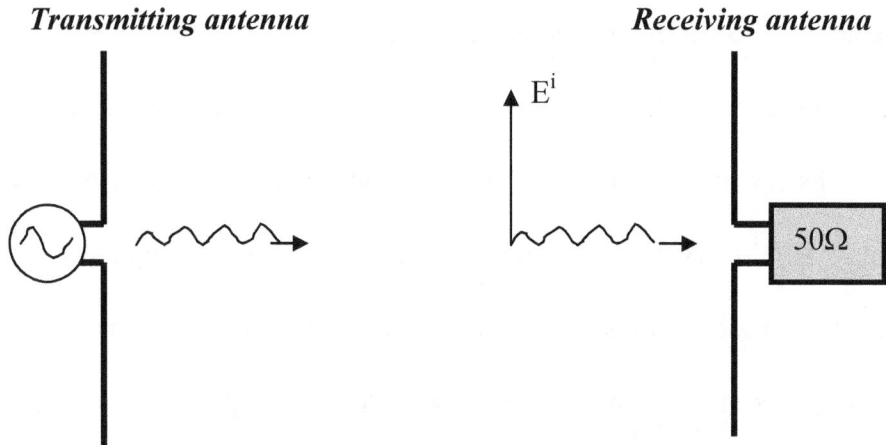

Figure 1.1 Complete Communication System including transmitting and receiving antenna.

1.2 Theory

1.2.1 Estimation of Voltage Waveform of Radiated Signal

For the transmitting case, a wire antenna is considered to be placed in a linear, isotropic and homogeneous dielectric medium. The antenna is excited by a wide band signal V(ω) of non-sinusoidal waveform. The time–domain plot of the input signal is given as follows

$$V(t) = F^{-1}[V(\omega)] \tag{1.1}$$

Here F^{-1} stands for the Inverse Fourier Transform.

The Inverse Fourier Transform of a continuous signal $V(\omega)$ is expressed as follows

$$V(t) = \frac{1}{2\pi} \int_{-\infty}^{\infty} V(\omega)\, e^{j\omega t}\, d\omega$$

(1.1a)

However, generally the signal is considered to have some discrete complex values $V_1, V_2, V_3, \ldots \ldots V_k \ldots \ldots V_F$ at discrete sampling frequency points (say F) in the domain covering the desired frequency band. Here V is a complex number

$$V_i = V_{real} + jV_{imag}$$

(1.2)

The frequency points are written as $f(n) = \Delta f\, n$; n=1,............, F and Δf is the frequency interval.

To apply the inverse discrete Fourier Transform, the signal must be defined at the corresponding negative frequencies as well, i.e. at n= -F,....,-1. The value at a negative frequency, $f(-n)$, is the complex conjugate of the corresponding value at the positive frequency, $f(n)$. The next step is to

5

convert the frequency sequence from $-F$ to F to the sequence 0 to 2 F, which is performed by using the following formula

$$V(n) = V*(N+1-n), \quad n = F+1,\ldots\ldots,N, \qquad N = 2F$$

(1.3)

The inverse Discrete Fourier Transform of a voltage pulse $V(n)$ is written as follows

$$V(k) = \frac{1}{N} \sum_{n=0}^{N-1} V(n) \, \exp\left(\frac{+j2\pi \, k \, n}{N}\right)$$

(1.4)

Here $V(n)$ are N values of the frequency spectrum of a pulse at frequencies $f(n)$ and $V(k)$ are N values of the time domain pulse form sampled at times t-Tk. The sampling time interval T is written as follows

$$T \equiv \frac{1}{\Delta f \, N}$$

(1.5)

Now, as already discussed, the induced current distribution on the wire due to a Delta-gap generator (for transmitting antenna) or plane wave incidence (receiving antenna), which is wideband in frequency domain, can be determined using Method of Moments with pulse basis function and point matching technique. The corresponding near field and far field components can be evaluated using the formulation presented and used in the previous chapters. The radiated far field due to the dipole divided into N current elements $I(n)\Delta L_n$ is expressed as a function of frequency as follows

$$E_\theta(\omega) = j\eta \frac{ke^{-jkr} \sin\theta}{4\pi r} \sum_{n=1}^{N} I_\omega(z')\Delta L_n e^{jkz' \cos\theta}$$

(1.6)

$E_\theta(\omega)$ in a particular θ direction is complex containing an amplitude and phase part. The variation of phase with frequency is highly nonlinear with frequency.

The time domain plot of the radiated electric field is found by taking the inverse Fourier Transform as follows

$$E_\theta(t) = F^{-1}[E_\theta(\omega)] \tag{1.7}$$

1.2.2 Estimation of Voltage Waveform of Received Signal

When an antenna receives a plane wave of angular frequency ω, the Complex Antenna Factor (CAF) is defined as follows

$$CAF(\omega) = \frac{Incident\ electric\ field\ (E^i\ (\omega))}{Re\ ceived\ voltage\ (V_R(\omega))} \tag{1.8}$$

Here E^i is the complex incident electric field with the polarization giving maximum output and V is the complex output voltage developed across the 50 Ω load.

By knowing the CAF of the antenna, the received voltage is evaluated from equation (1.4) as follows

$$V_R(\omega) = \frac{E^i(\omega)}{CAF(\omega)} \tag{1.9}$$

The derivation for scalar antenna factor has already been described in Chapter 3. Here, some of the related expressions have been repeated for the sake of completeness.

To evaluate the received voltage theoretically, the sensor is replaced by a open circuit voltage V_{oc} connected with the output impedance Z_{out} of the sensor. The voltage to the receiver (Figure 1.2) is achieved as follows

7

$$V_R(\omega) = \frac{Z_L}{Z_L + Z_{out}} V_{oc}(\omega)$$

(1.10)

Generally Z_L i.e. impedance of the detector (e.g. spectrum analyser) is considered as 50 ohm. The open-circuited voltage is evaluated using the following relation

$$V_{oc}(\omega) = \vec{E}^i(\omega)\ \vec{l}_{effective}(\omega) \tag{1.11}$$

Here the effective length is considered as a complex quantity i.e. the phase information is also taken into account .

It is noticed from equation (1.11) that the open-circuited voltage is a complex quantity and hence the antenna factor is also a complex quantity and its phase is determined by the phase of the output voltage. The phase is not constant or linear with frequency but it is highly nonlinear with frequency.

The time domain spectrum of the output signal is given as follows

$$V_R(t) = F^{-1}[V_R(\omega)]$$

(1.12)

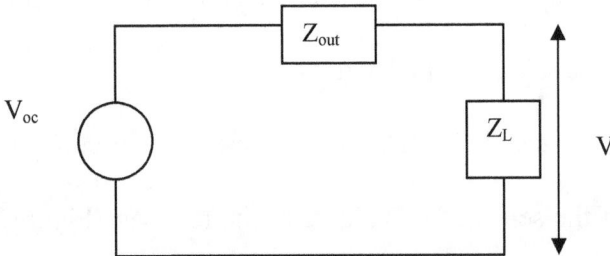

Figure 1.2 Equivalent circuit diagram of a sensor.

1.3 Results

In this chapter, the phase vs. frequency characteristics of the signal transmitted /received by loaded (e.g. inverted L, T, C and I-shaped antenna) and unloaded wire antennas are studied.

Next, a numerical study is made on the performance of a dipole as transmitting and receiving antenna when a wideband signal is applied to the input of the transmitting antenna. A dipole antenna of length 0.15m and radius=0.2mm is excited by a signal of amplitude 1Volt which is broadband in frequency domain (0.4 GHz-2.2 GHz) (Figure 1.15(a)) but an impulse in the time domain (Figure 1.15(b)). The time domain spectrum of the input signal is achieved by using the Inverse Fourier Transform of the frequency-domain pulse form. The signal radiated by the dipole is received by another identical receiving antenna, connected with a load of 50Ω at the gap and placed at a distance of 1m from the transmitting antenna on its axis. The transmitted electric field at the position of the receiving antenna is evaluated considering the receiving antenna as absent. The frequency spectrum of the dominant component of the radiated electric field at a distance of 1m from the transmitting antenna in the absence of the receiving antenna is shown in Figure 1.16. This transmitted signal acts as the excitation for the receiving antenna. The frequency spectrum of the output voltage received at the load terminal is shown in Figure 1.17(a). The time domain plot of the output signal is achieved by taking the inverse Fourier Transform of the frequency-domain data and is shown in Figure 1.17(b).

Voltage V

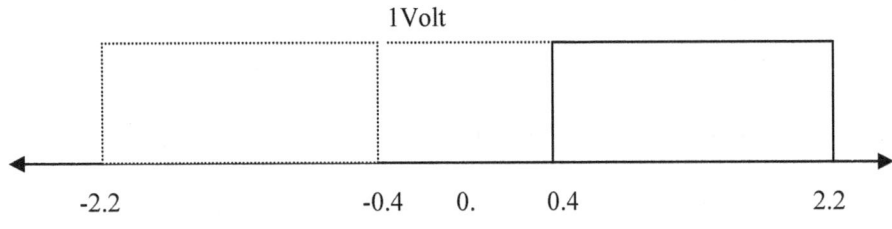

1Volt

-2.2   -0.4   0.   0.4   2.2

Frequency in GHz

Figure 1.15(a) Plot of input signal in the frequency domain.

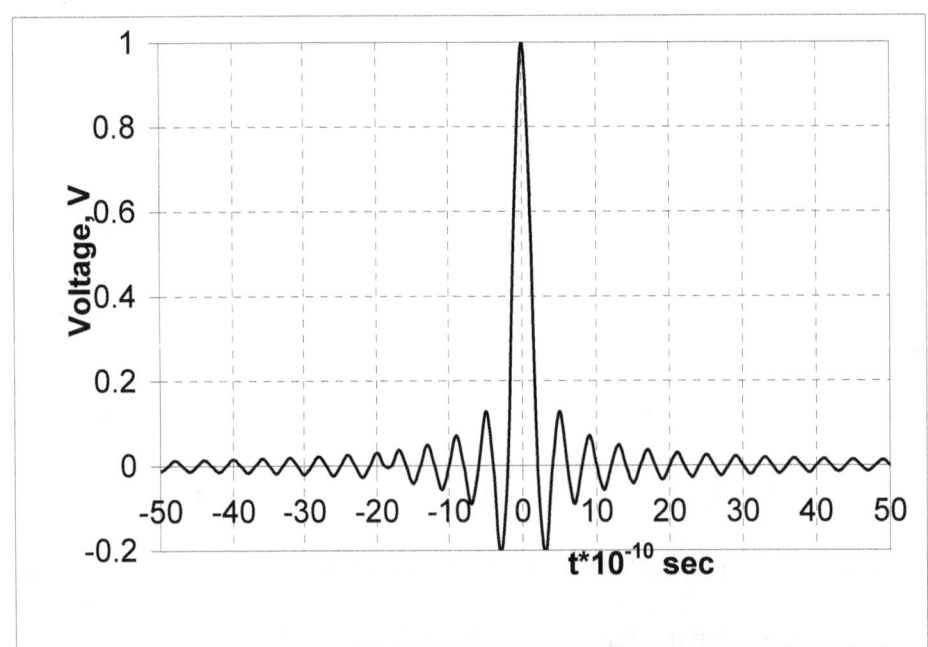

Figure 1.15(b) Input voltage waveform in time domain.

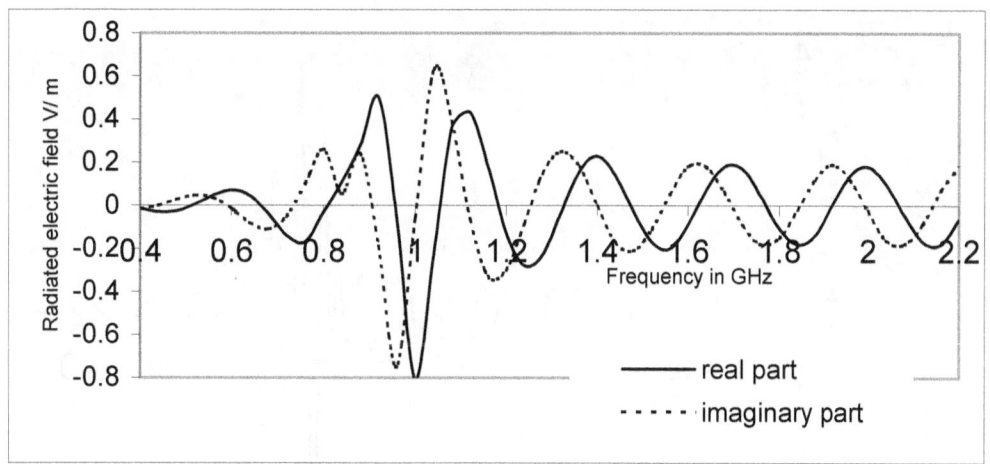

Figure 1.16 Dominant component of radiated electric field as a function of
frequency at a distance of 1m in the broadside direction from the
transmitting antenna.

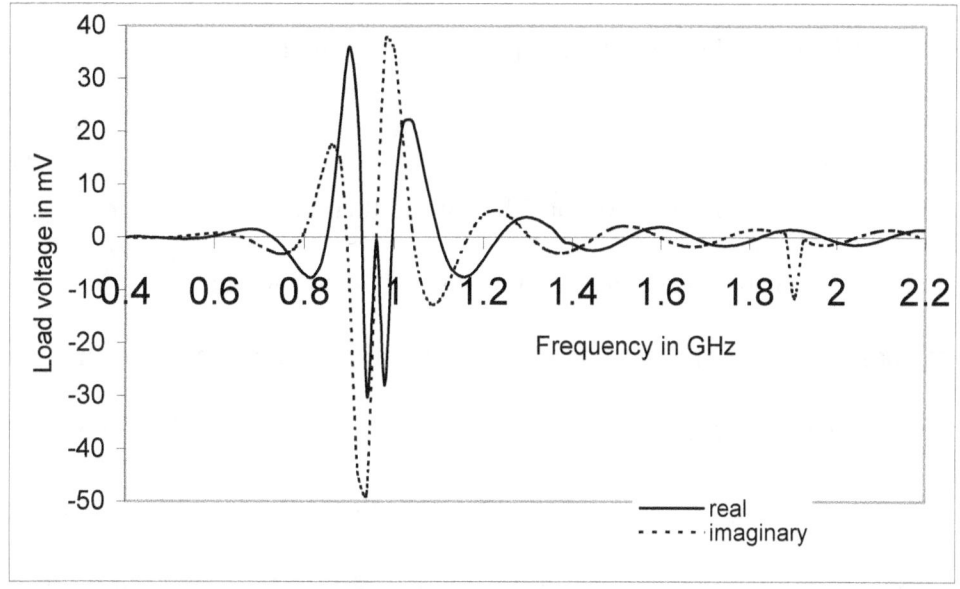

Figure 1.17(a) Plot of the output voltage developed at the load of the
receiving antenna in frequency domain.

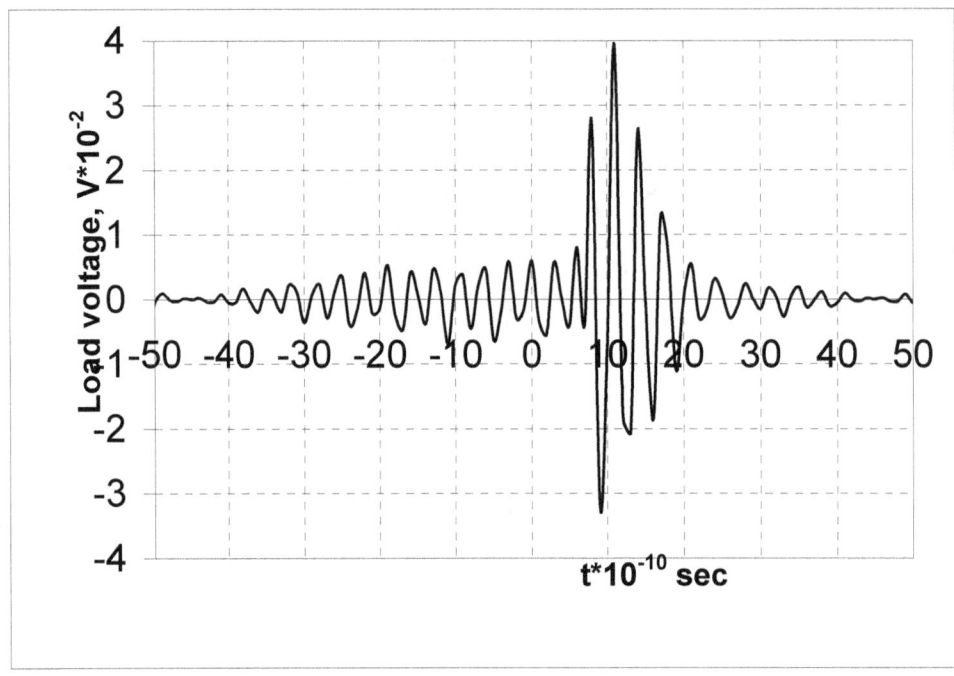

Figure 1.17(b) Plot of output voltage in time domain.

Finally the results have been validated with the theoretical results for the generator and output voltage waveform of the receiving antenna in the time domain, reported by David M. Pozar . Pozar had optimized the generator voltage waveform and presented the corresponding output voltage waveform in the time domain for a pair of lossless dipoles each of length L=15 cm, radius=0.02cm, $\sigma = \infty$, generator impedance Z_G=50 ohms and load Impedance $Z_L = \infty$. The dipoles are assumed as parallel, radiating in the broadside directions and placed in the far field region of the other over the operating bandwidth. In this thesis, for the sake of comparison, the same optimization formulation is used to evaluate the generator voltage in the frequency domain and the corresponding plot in the time domain is shown in Figure 1.18(a). The time domain plot of the received voltage for the antenna parameters given above is presented in Figure 1.18(b).

For the evaluation of the transfer function, the input voltage is taken as 1volt.

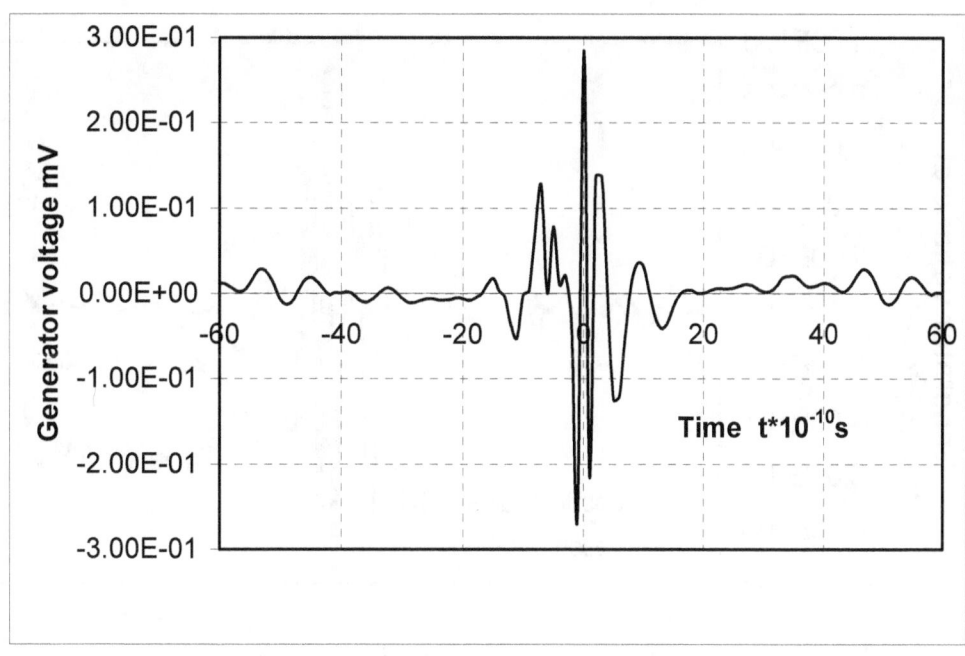

Figure 1.18(a) Time domain plot of generator voltage.

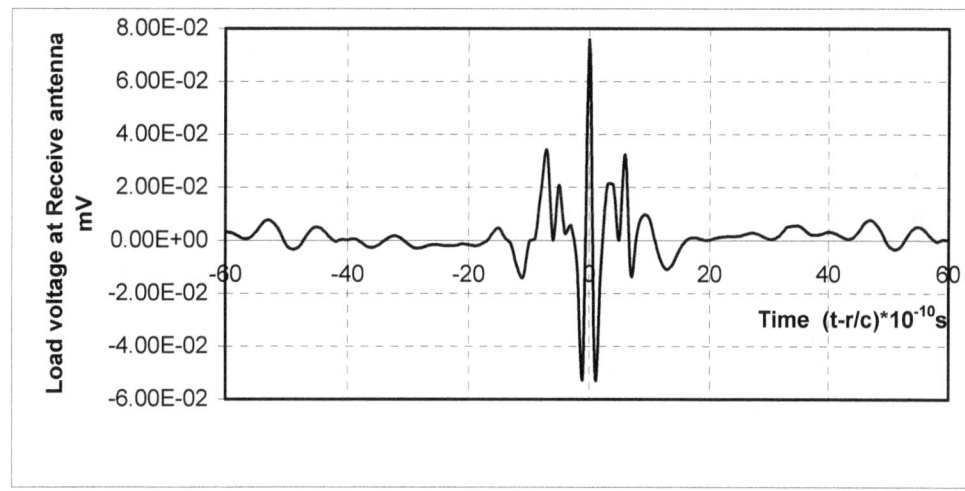

Figure 1.18(b) Time domain plot of the voltage developed at the load of the receiving antenna.

The results have been put in the same scale to compare with the reported value.

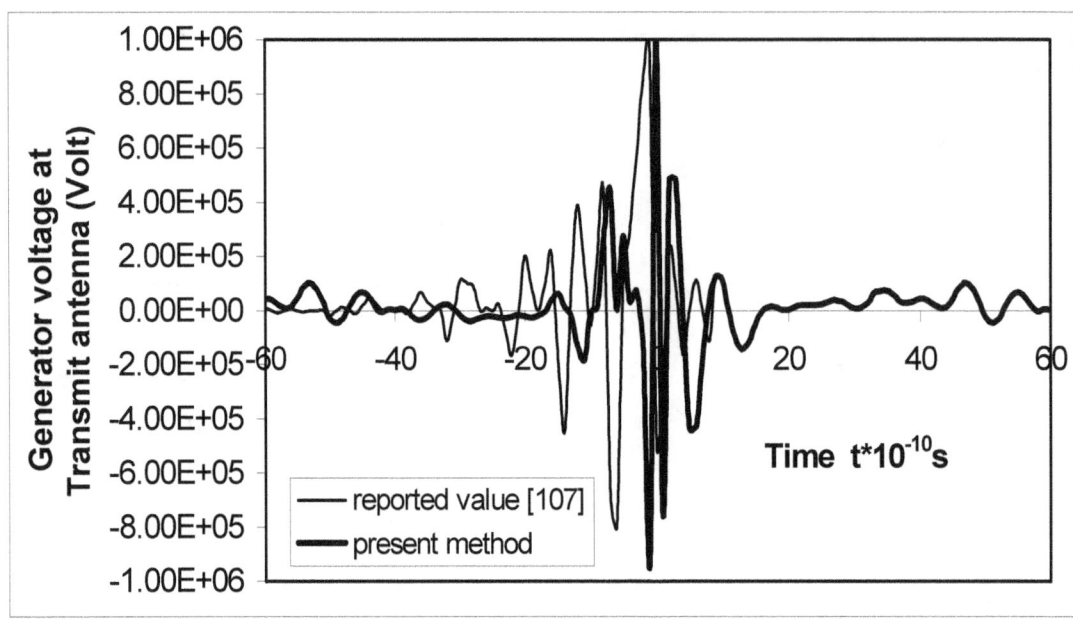

Figure 1.19(a) Generator voltage waveform in time domain.

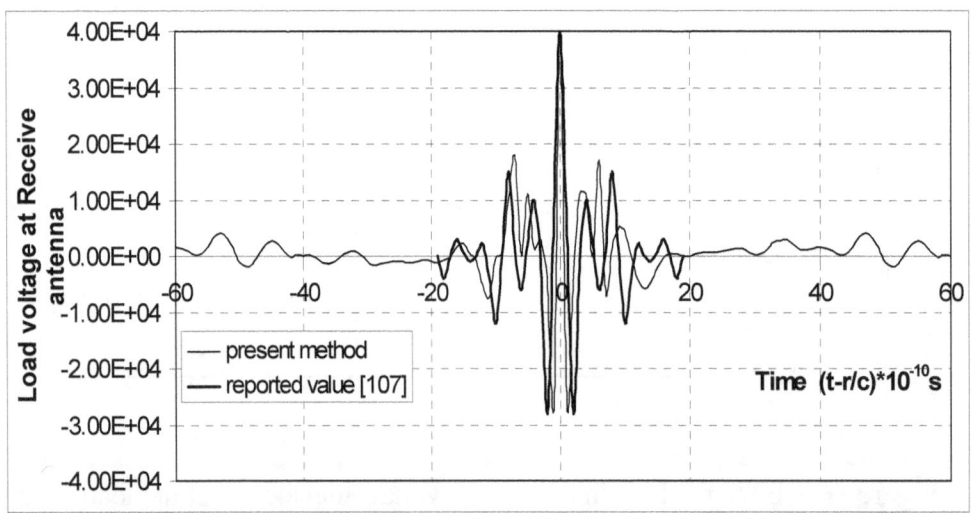

Figure1.19(b) Receive antenna voltage waveform in the time domain.

1.4 Discussions

From the phase versus frequency plot of the transmitted and received signal, it is noticed that for most of the cases, the response is highly non-linear with frequency. As an example, the phase versus frequency plot of the signal transmitted by a dipole antenna (Figure 1.3(a)) is considered. The nonlinearity in the phase variation with frequency proves the dispersive nature of the far-field radiated signals. But, however, in a selected narrow band region, we can have approximately linear characteristics. In Figure 1.3(a), a linearised phase variation with negative slope is noticed in the region 1GHz-1.2GHz. This negative slope means a positive delay in the time response. Similarly, we have a literalized phase variation of positive slope in the frequency band 1.2GHz - 1.35GHz. Also in the frequency band 0.5 GHz – 0.8GHz the phase is almost constant and hence the dispersion is minimum.

Similar studies have been performed for loaded antennas. For inverted L-shaped antenna (Figure 1.4(a)) the following characteristics are noticed:

i)   linearised phase variation with negative slope in the range 0.4 - 0.6 GHz

ii)   linearised phase variation with positive slope in the range 0.6 - 0.8 GHz

iii)   almost constant phase in the range 0.9GHz - 1.3 GHz.

For T-shaped antenna (Figure 1.5(a)) the following characteristics are noticed:

i)   linearised phase variation with negative slope in the range 0.6 - 0.8 GHz

ii)   linearised phase variation with positive slope in the range 0.8 - 1.0 GHz

iii)   almost constant phase in the range 1.5GHz - 2.0 GHz.

For I-shaped antenna (Figure 1.6(a)) the phase remains almost constant in the frequency range 0.5GHz-1.0GHz. Similarly for C-shaped antenna the phase versus frequency plot shows a flat region in the frequency range 1 GHz – 2 GHz.

For receiving case also, certain frequency regions are observed with almost constant phase – i) 0.5GHz – 0.9GHz for dipole antenna (Figure 1.3(b)), ii) 0.7GHz – 1.6GHz for inverted L antenna (Figure 1.4(b)), iii) 1GHz – 1.6GHz for T-shaped antenna (Figure 1.5(b)), iv) 0.4GHz – 1GHz for I-shaped antenna (Figure 1.6(b)), v) 1GHz – 1.5GHz for C-shaped antenna (Figure 1.7(b)).

Hence from the above studies of the phase versus frequency plots, it has been realised that the loaded and unloaded transmitting / receiving antennas show non-dispersive characteristics within a certain frequency band i.e. act like a frequency selective network. These plots play important role to achieve the distortion in the output waveform while working with a broadband signal.

Also from the plot of Complex Antenna Factor versus frequency for the loaded and unloaded antennas, certain frequency regions with low antenna factor and almost constant phase are achieved. – i) 0.4GHz – 0.9GHz for dipole antenna (Figure 9.8), ii) 0.7GHz – 1.6GHz for inverted L antenna (Figure 1.9), iii) 1GHz – 1.6GHz for T-shaped antenna (Figure 1.10), iv) 0.4GHz – 1GHz for I-shaped antenna (Figure 1.11), v) 1GHz – 1.5GHz for C-shaped antenna (Figure 1.12), vi) no such region is observed for folded dipole (Figure 1.13) and vii) 0.8GHz – 1.2 GHz for loop antenna (Figure 1.14). Hence, in the present situation, it has been found that the antenna is acting like a frequency selective network or like a filter of some form both in transmitting and receiving mode.

To study the distortion in the output waveform, an antenna is considered to be excited by a wideband signal and received by another receiving antenna of same dimension. The time-domain plot (Figure 1.15(b)) of the input pulse signal (Figure 1.15(a)) looks like a narrow impulse. But the plots of the radiated electric field (Figure 1.16) and received voltage at the load versus frequency (Figure 1.17(a)) show that, the antenna both in transmitting and receiving mode, have dispersive characteristics. Also the received voltage in the time domain (Figure 1.17(b)) is not an impulse function as the input signal. Here the receiving antenna is placed in the near field region of the transmitting antenna. Thus the present theory can be used for far field as well as for near field.

Next, for the purpose of validation, the results have been generated for two identical dipoles of the given dimension. The results for the generator and output voltage in the time domain (Figure 1.18(a)-(b)) show that the trend of the graphs is same. However, a very little shift in the time scale is noticed for the generator voltage plot for the two cases. This is mostly due to the fact that, in the literature, each antenna is considered to be placed in the far field region of the other, (though the exact value is not given), whereas in the present work, the near field effect with the mutual coupling is considered.

1.5 Conclusion

From the studies of this chapter, it can be concluded that whenever there is a requirement of using ultra wide band signal, one has to realize linear phase versus frequency response of the system and constant amplitude in the frequency spectrum both in transmitting and receiving mode. However the loaded and unloaded antennas act like a frequency selective network or a filter of some form both in transmitting and receiving mode.

The study in this chapter on the phase vs. frequency characteristics plays an important role and knowledge of the same is extremely essential for dealing with practical waveforms, which are non-sinusoidal and wide-band in nature.

Chapter-2

Trans-Receive Characteristics of the Monopole Antenna Using WIPL-D:

The transmitting and receiving antennas are considered on an infinite ground plane. The transmitting antenna is fed by a 1V voltage source and receiving antenna is terminated by a 50 ohm load.

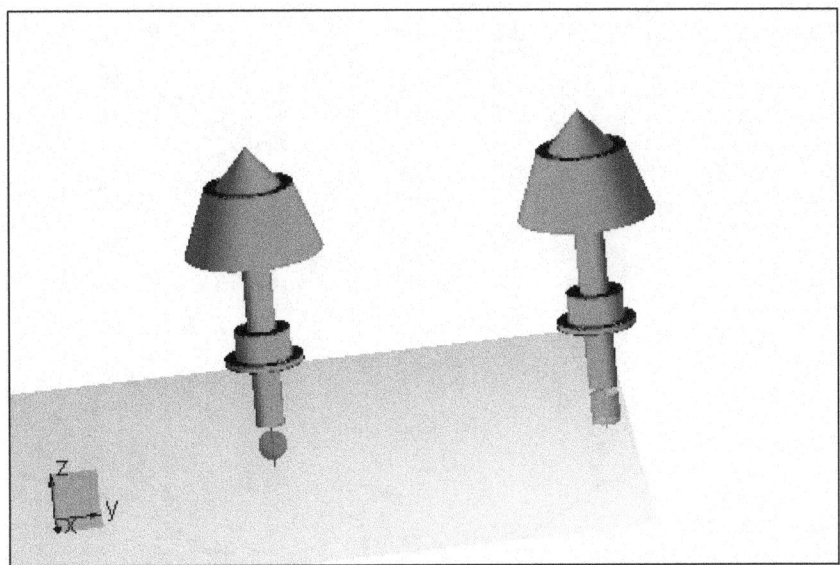

Figure 2.1: The transmitting and receiving antennas on an infinite ground plane.

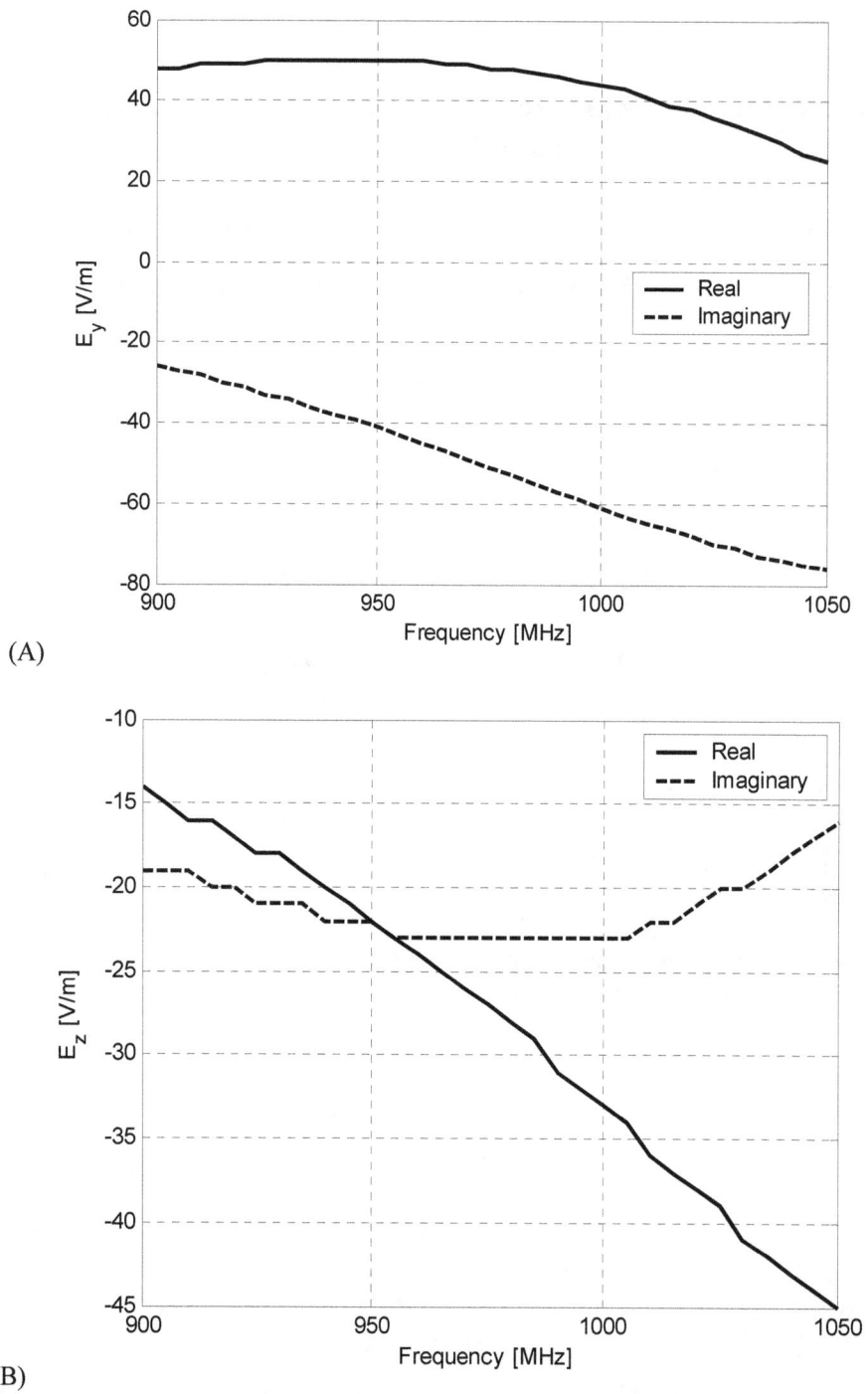

(A)

(B)

Figure 2.2. Simulated incident electric field at a point x = 0, y = 30, z = 29 mm (near field) (A) E_y (B) E_z component.

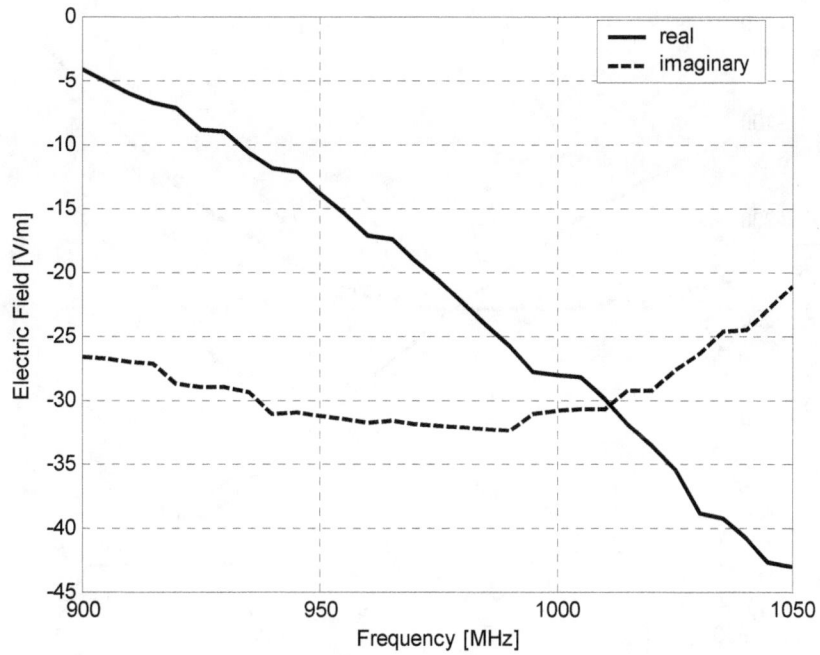

Figure 2.3. E$_\theta$ component of electric field at a point x = 0, y = 60, z = 29 mm (far field).

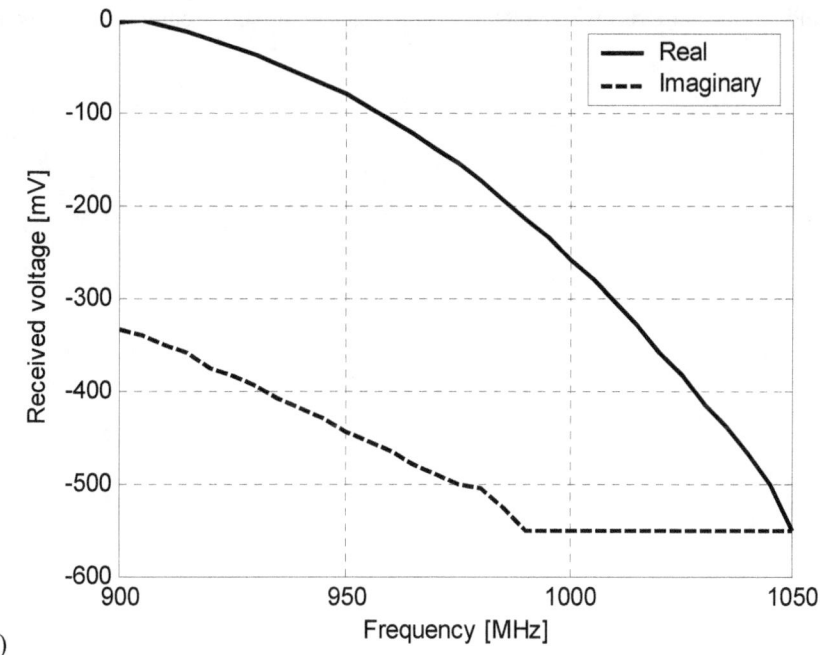

(A)

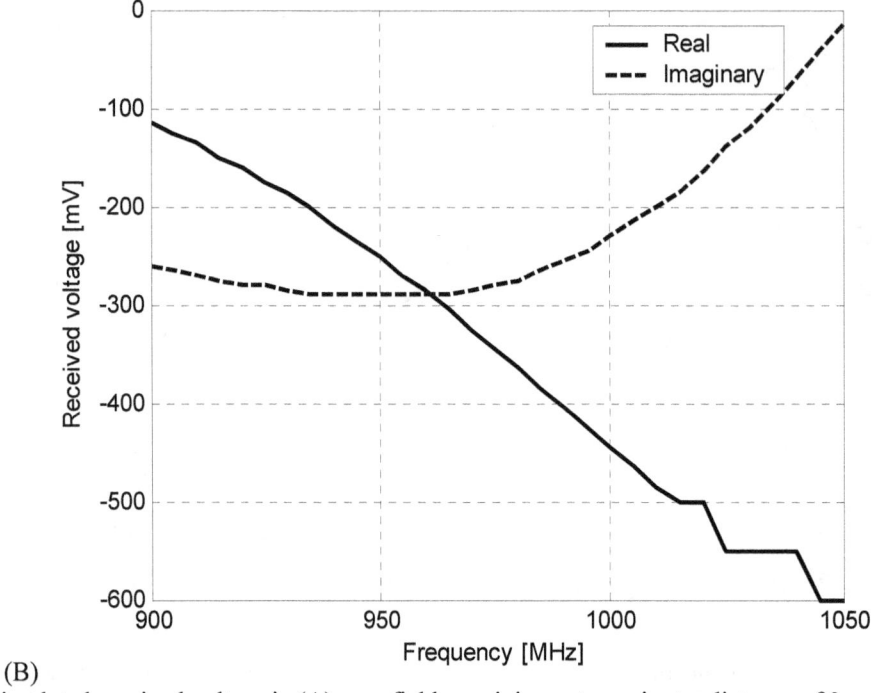

(B)

Figure 2.3: Simulated received voltage in (A) near field, receiving antenna is at a distance of 3 cm from transmitting antenna and (B) far field, receiving antenna is at a distance of 6 cm from transmitting antenna.

Chapter 3

In this chapter we will consider the issues related to designing MATCHED filter for LFM pulse. Although Matched filter can counter the effect of ANTENNA TRANSFER FUNTION after redesign, it can not compensate Channel transfer function. For Channel transfer function and added NOISE we have to incorporate suitable adaptive Equalizer along with matched filter. In this chapter we will describe the design of ADAPTIVE EQUALIZER with three sample channel transfer function after redesigning of MATCHED Filter.

Designing MATCHD FILTER for LFM RADAR

A filter is said to be matched to the signal x (t) if the filter impulse response is h(t) = kx(-t+t$_0$)
$$=k \, x[-(t- t_0)]$$
The impulse response takes the shape of x(t) delayed by t$_0$ and reversed in time.
Here k & t$_0$ are const.
The principle of the match filter:
We want to maximize the o/p signal component and minimize the noise component in a small frequency band Δf. Near the frequencies (1/ Δ t) i.e. the zero amplitude point of the spectrum, the signal amplitude are small and contribute a negligible amount to the output-signal power. So the gain of the filter should be quite small for this frequency component. We conclude that the pass band characteristics matching the frequency spectrum of the signal would provide maximum output signal and minimum noise.
For basis network analysis:
The network has two terminals, linear and fixed parameter, the parameters do not vary with time.
Transfer function: the transfer function of linear network H (f) is defined as the ratio of o/p to i/p in frequency domain.
$$H (f) = Y (f)/ X (f).$$
$$\Rightarrow \; |Y(f)| = |H(F)| X(f)$$

Unit Impulse Response: The unit impulse response h (t) of the network is the o/p function for an i/p of unit impulse function, i.e.
$$x (t) = \delta(t), \quad y(t) = h(t).$$
Understanding of network behavior, the unit impulse response h(t) is very important.
The relation between the unit impulse response h(t) & the transfer function H (f) of the network is

$$h(t) = \int_{-\infty}^{+\infty} H(f)e^{j\omega t} df$$

Conversely, H (f) = $\int_{-\infty}^{+\infty} h(t)e^{-j\omega t} dt$

Convolution Integral: Output Signal in Time Domain
Convolution theorem: convolution in the time domain corresponds to multiplication in frequency domain, and vice versa.

$$x(t) * y(t) \leftrightarrow X(f)\ Y(f)$$
$$x(t)\ y(t) \leftrightarrow X(f) * Y(f)$$

The Fourier transform method provides input output relationship in the frequency domain. To obtain the relationship in the time domain, we would use convolution integration.

The advantage of convolution integral is that it provides a direct methods of obtaining the output signal from the input signal x (t) and impulse response h (t) of the network.

We get, $y(t) = \int\limits_{0}^{\infty} h(\tau)x(t-\tau)d\tau$

$$y(t) = \int\limits_{-\infty}^{t} x(\tau)h(t-\tau)d\tau.$$

Network with Random Input:

The relationships between input and output using convolution integral for random input signals to a network.

The random process can be completely described by three speciation .they are

1) Autocorrelation function.
2) Spectral density.
3) Probability density function.

Their use in determining input/output relationship for a random process.

Autocorrelation input-output relationship:

Let y(t) the output of the network, be a sample function of random process. its autocorrelation function is defined as the time average of $y(t_1)y(t_2)$.

$R_y(\tau) = \langle y(t_1)\ y(t_2) \rangle$ where $t_2-t_1 = \tau$.

$$y(t_1) = \int\limits_{-\infty}^{+\infty} h(\tau_1)x(t_1-\tau_1)d\tau_1$$

$$y(t_2) = \int\limits_{-\infty}^{+\infty} h(\tau_2)x(t_2-\tau_2)d\tau_2$$

We get $R_y(\tau) = \left\langle \int\limits_{-\infty}^{+\infty} h(\tau_1)x(t_1-\tau_1)d\tau_1 \int\limits_{-\infty}^{+\infty} h(\tau_2)x(t_2-\tau_2)d\tau_2 \right\rangle$

$$= \int\limits_{-\infty}^{+\infty} h(\tau_1) \int\limits_{-\infty}^{+\infty} h(\tau_2)\langle x(t_1-\tau_1)x(t_2-\tau_2)\rangle\ d\tau_1 d\tau_2$$

Clearly, $\langle x(t_1-\tau_1)\ x(t_2-\tau_2)\rangle = R_x (t_2-t_1-\tau_2+\tau_2) = R_x (\tau-\tau_2+\tau_2)$

Then $R_y(\tau) = \int\limits_{-\infty}^{+\infty} h(\tau_1) \int\limits_{-\infty}^{+\infty} h(\tau_2)R_x(\tau-\tau_2+\tau_1)d\tau_1 d\tau_2$

This result shows the o/p autocorrelation function in term of i/p for a random process.

Spectral   Density Input Output Relationship:

$$G_y(f) = \int_{-\infty}^{+\infty} R_y(\tau)e^{-j\omega\tau}d\tau$$

Put the value
$$G_y(\tau) = \int_{-\infty}^{+\infty}[\int_{-\infty}^{+\infty} h(\tau_1)d\tau_1 \int_{-\infty}^{+\infty} h(\tau_2)R_x(\tau-\tau_2+\tau_1)e^{-j\omega\tau}d\tau_2]d\tau$$

$$= \int_{-\infty}^{+\infty}[\int_{-\infty}^{+\infty} h(\tau_1)d\tau_1 \int_{-\infty}^{+\infty} h(\tau_2)d\tau_2 \int_{-\infty}^{+\infty} R_x(\tau-\tau_2+\tau_1)e^{-j\omega\tau}]d\tau$$

We multiply the right hand side by $e^{j\omega\tau_1}e^{-j\omega\tau_2}e^{-j\omega(\tau_1-\tau_2)}(=1)$

Therefore

$$G_y(f) = \int_{-\infty}^{+\infty} \int_{-\infty}^{+\infty} h(\tau_1)e^{j\omega\tau_1}d\tau_1 \int_{-\infty}^{+\infty} h(\tau_2)e^{-j\omega\tau_2}d\tau_2 \int_{-\infty}^{+\infty} R_x(\tau-\tau_2+\tau_1)e^{-j\omega(\tau-\tau_2+\tau_1)}d\tau$$

Now let us identify each integral:

$$\int_{-\infty}^{+\infty} h(\tau_1)e^{j\omega\tau_1}d\tau_1 = H(-f)$$

And $\int_{-\infty}^{+\infty} h(\tau_2)e^{-j\omega\tau_2}d\tau_2 = H(f)$

For the third integral, by getting $\tau' = \tau - \tau_2 + \tau_1$, we get

$\int_{-\infty}^{+\infty} R_x(\tau')e^{-j\omega\tau'}d\tau'$ Which is equal to $G_x(f)$.

Then $G_y(f) =$ H (f). H (-f). G $_x$(f).

$\qquad = |H(f)|^2 \, G_x(f).$

We observe:
1) The calculation for the spectral density are less difficult than autocorrelation calculations. So we conclude that analysis in frequency domain is generally preferred than time domain when dealing with random process.
2) If the impulse response of the network specified, the o/p spectral density or autocorrelation function can be specified.

Signal Detection:

Correlation between two signals is an extremely important concept, which measures the degree of similarity (agreement or alignment) between the two signals. This is widely used for signal processing application (in radar, sonar, digital communication, electronics warfare, and many others.

The case of binary communication, the two known wave form are received in a random sequence. We must take the two pulses dissimilar as possible. we should select highest dissimilarity (c_n = -1). This scheme is sometimes called the ***antipodal*** scheme. We can use ***orthogonal*** pulses, which result in c_n =0. In practice both this option are used, but the antipodal one is best terms of distinguish ability between the two pulses. For antipodal scheme in which the two pulses are p (t) and –p (t).there is always an unwanted signal (noise) superimposed on the received pulses. Then pulses are distorted.   The c_n value is no longer±1. For distinguishibility, we use a **threshold detector**, it decides that if the correlation is positive [p(t)] or negative[-p(t)].

For threshold detection method, the received pulses are sampled as its peak amplitude A_P. But for channel noise, the sampled value (A_P +n). The decision is made from the value (A_P +n).

The received pulse p(t) be time limited to T_0. By passing the received pulses through a filter that enhances the pulse amplitude at some instant t_m and simultaneously the reduces the noise power. We thus seek a filter with a transfer function H (f) that maximizes the (SNR).

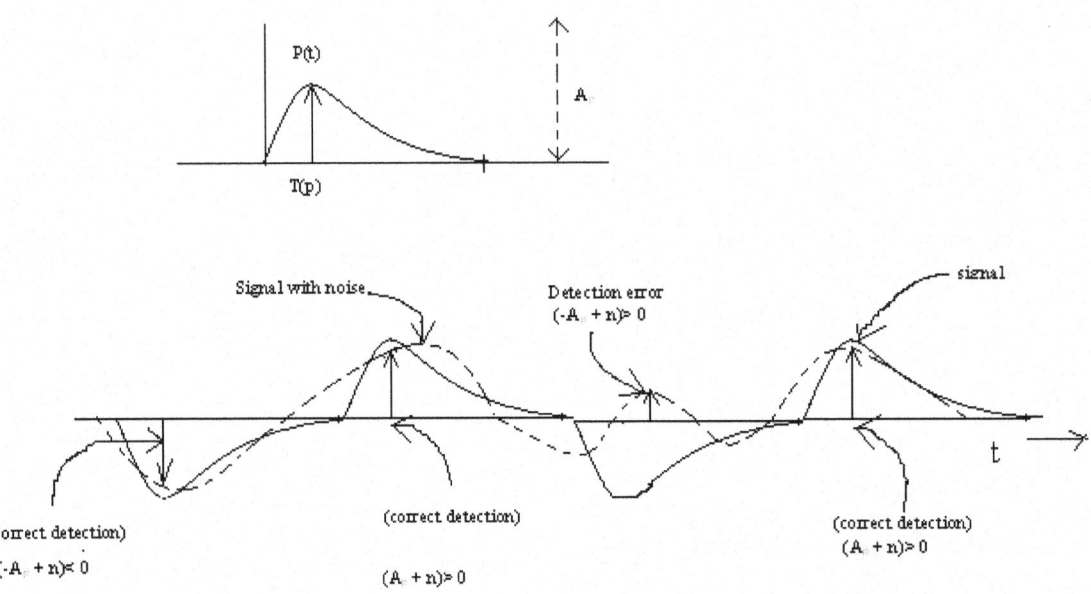

Error probability in threshold detection

Match Filter
A filter is matched to the signal x(t) if the FILTER'S Unit Impulse Response is h(t) = kx(-t + t0) = kx[-(t-t0)] , get maximizes output signal to noise ratio(SNR).

Let $t_1 = -t + t_0$

Now $X(-j\omega) = \int\limits_{-\infty}^{\infty} x(t_1)e^{+j\omega t_1} dt_1$

Now $X^*(j\omega) = \int\limits_{-\infty}^{\infty} x(t_1)^* e^{+j\omega t_1} dt_1$

Since x(t) is a real signal $x^*(t_1) = x(-t_1)$,hence $X^*(j\omega) = X(-j\omega)$]

Hence Transfer Function $H(j\omega) = \int\limits_{-\infty}^{\infty} h(t)e^{-j\omega t}\, dt$

$$= k\int\limits_{-\infty}^{\infty} x(-t+t_0)e^{-j\omega t}\, dt$$

$$= ke^{-j\omega t_0}\int\limits_{-\infty}^{\infty} x(t_1)e^{-j2\pi(-f)t_1}\, dt_1$$

$$= ke^{-j\omega t_0} X(-j\omega)$$

$$= ke^{-j\omega t_0} X^*(j\omega)$$

$$\therefore H(j\omega) = ke^{-j\omega t_0} X^*(j\omega)$$

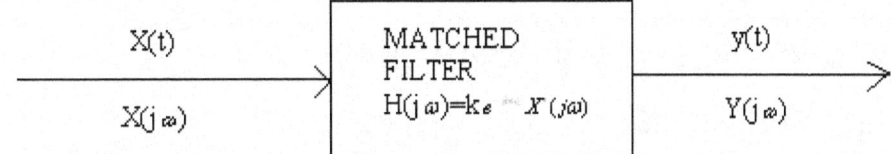

$$Y(j\omega) = H(j\omega)\, X(j\omega)$$
$$= ke^{-j\omega t_0} X(j\omega)\, X^*(j\omega)$$
$$= k\left|X(j\omega)\right|^2 e^{-j\omega t_0}$$
$$= k\, S_x(j\omega)\, e^{-j\omega t_0}$$
$$S_x(j\omega) = \left|X(j\omega)\right|^2 = \text{E.S.D. of SIGNAL}$$

Hence $y(t) = F^{-1}[Y(j\omega)] = k\int\limits_{-\infty}^{\infty} S_x(j\omega)\, e^{+j\omega(t-t_0)}\, df$

$$= k\int\limits_{-\infty}^{\infty} S_x(j\omega)\cos\omega(t-t_0)df + jk\int\limits_{-\infty}^{\infty} S_x(j\omega)\sin\omega(t-t_0)df$$

$$= k\int\limits_{-\infty}^{\infty} S_x(j\omega)\cos\omega(t-t_0)df$$

$$\therefore y(t)_{MAX} = k\int\limits_{-\infty}^{\infty} S_x(j\omega)\, df = kE_x$$

[$S_x(j\omega)$ is an even function . The 2nd part of the integration is ZERO
As $S_x(j\omega) > 0$ the $y(t)$ will be maximum at $t = t_0$, E_x = total Energy of Signal]

For Time Varying NOISE (General Case)

At t= t_0 average output signal power $\overline{x^2(t)} = \left| \int\limits_{-\infty}^{\infty} H(j\omega)X(j\omega)e^{j\omega t_0} df \right|^2$

Average power of the output noise at $t - t_0 = \overline{n_0^{\,2}(t)} = \int\limits_{-\infty}^{\infty} |H(j\omega)|^2 S_n(j\omega) df$

$\therefore (S/N)_{OUT} = \left| \int\limits_{-\infty}^{\infty} H(j\omega)X(j\omega)e^{j\omega t_0} df \right|^2 \bigg/ \int\limits_{-\infty}^{\infty} |H(j\omega)|^2 S_n(j\omega) df$

Let $A(j\omega) = H(j\omega)\sqrt{S_n(j\omega)}$, $B(j\omega) = \dfrac{X(j\omega)e^{j\omega t_0}}{\sqrt{S_n(j\omega)}}$

We know Schularz inequality is –

$$\left| \int\limits_{-\infty}^{\infty} A(j\omega)B(j\omega) df \right|^2 \le \int\limits_{-\infty}^{\infty} |A(j\omega)|^2 df \int\limits_{-\infty}^{\infty} |B(j\omega)|^2 df$$

Equality is obtained only when $A(j\omega) = kB^*(j\omega)$

Hence applying above we get $(S/N)_{out} = \dfrac{\int\limits_{-\infty}^{\infty} |H(j\omega)|^2 S_n(j\omega) df \int\limits_{-\infty}^{\infty} \dfrac{|X(j\omega)|^2}{S_n(j\omega)} df}{\int\limits_{-\infty}^{\infty} |H(j\omega)|^2 S_n(j\omega) df}$

or, $(S/N)_{out} \le \int\limits_{-\infty}^{\infty} \dfrac{|X(j\omega)|^2}{S_n(j\omega)} df$

Hence maximum $(S/N)_{out}$ can obtain if $H(j\omega)$ is so chosen such that $A(j\omega) = k\,B^*(j\omega)$
or, maximum $(S/N)_{out}$

or, $H(j\omega)\sqrt{S_n(j\omega)} = \dfrac{kX^*(j\omega)e^{-j\omega t_0}}{\sqrt{S_n(j\omega)}}$

$\therefore H(j\omega) = \dfrac{kX^*(j\omega)e^{-j\omega t_0}}{S_n(j\omega)}$ For maximum $(S/N)_{out}$

28

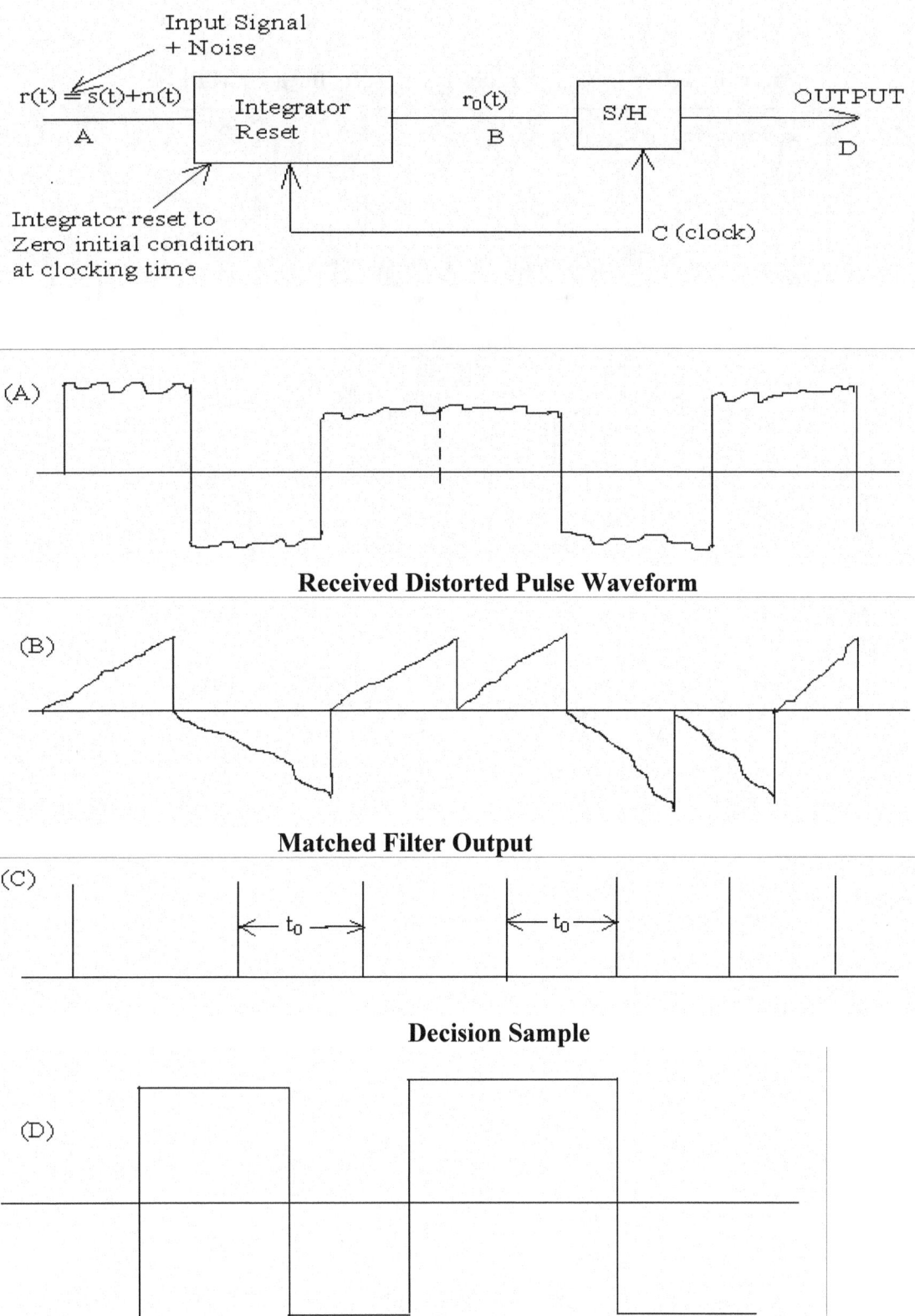

Received Distorted Pulse Waveform

Matched Filter Output

Decision Sample

Regenerated Pulse Waveform

Matched Filter Design: Simulation Results for a LFM Pulse

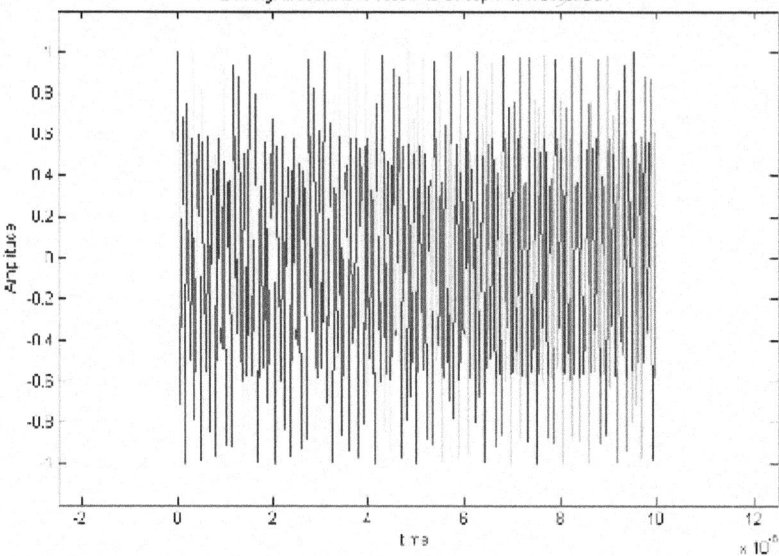

Fig.1 LFM signal (Chirp) pulse width 100nS, Sweep 150MHz
Centre Frequency 975MHz (900-1050MHz)

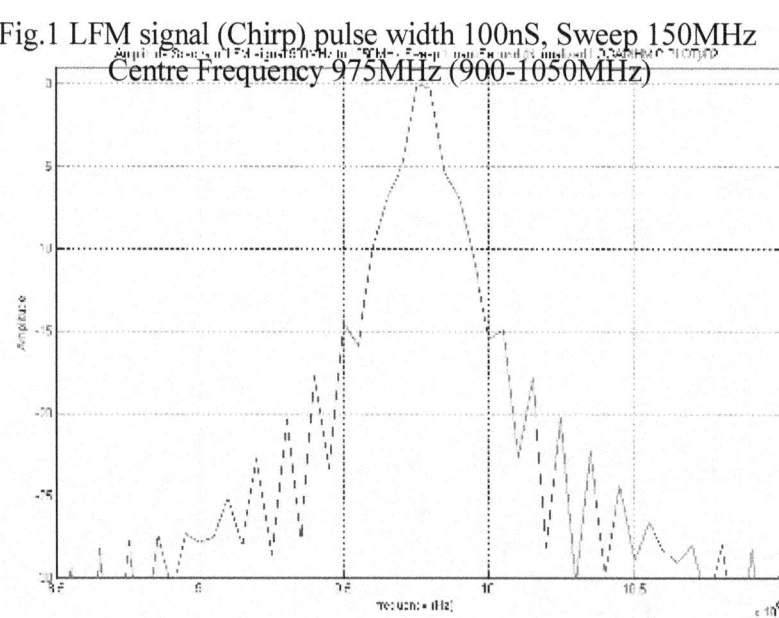

Fig.2 Amplitude Spectra of LFM pulse with Centre Frequency of 975MHz, Sweep 150MHz
(showing almost symmetry around Centre Frequency)

30

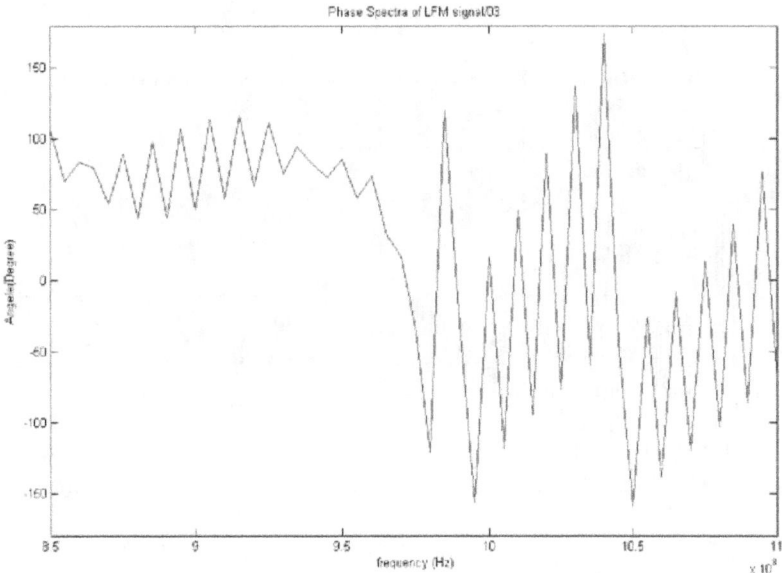

Fig.3 Phase Spectra of LFM Signal

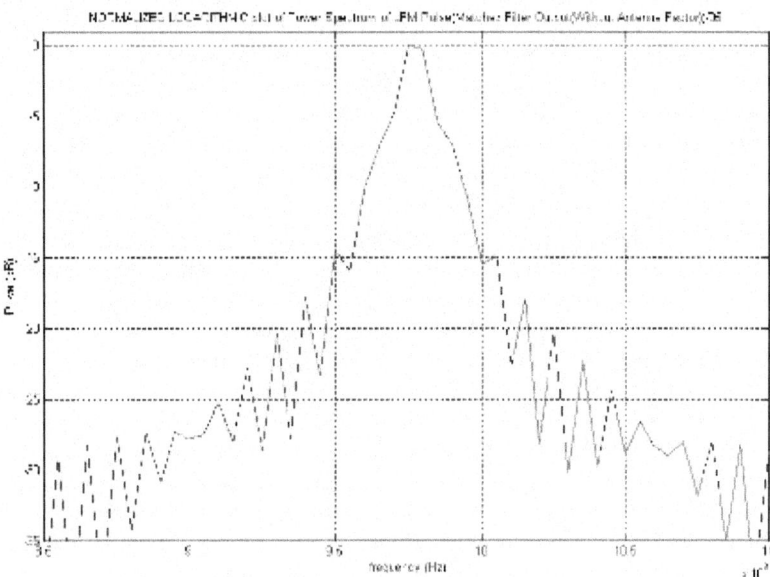

Fig.4 Without considering Antenna Transfer Function (ATF) into account Power spectra is almost replicate the Amplitude spectra

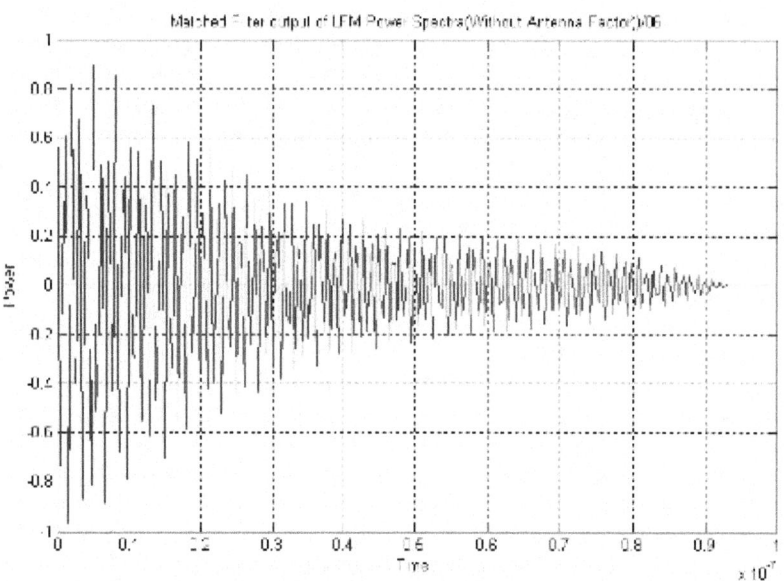

Fig.5 Time domain Matched Filter response shows that without considering ATF, we can obtained SINC pattern with peak at t=0(or at desired point)

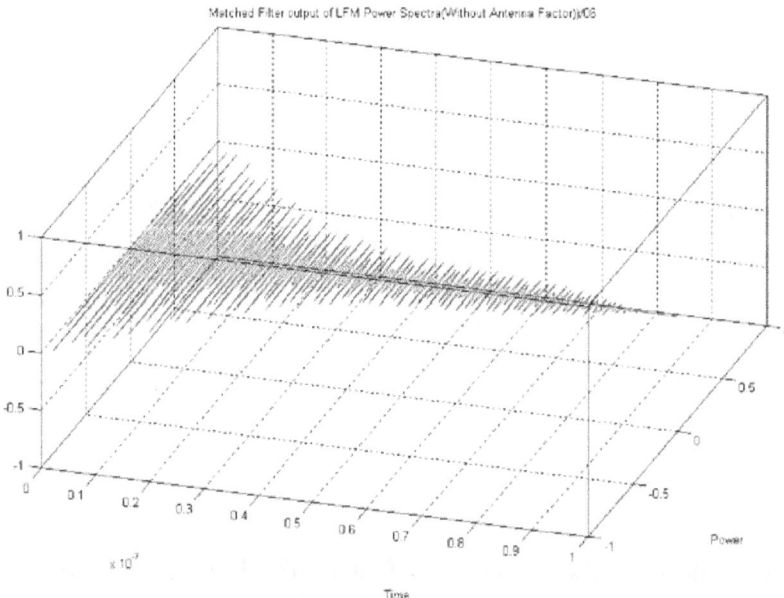

Fig.6 3D View of Matched Filter Response in time domain… (Without ATF)

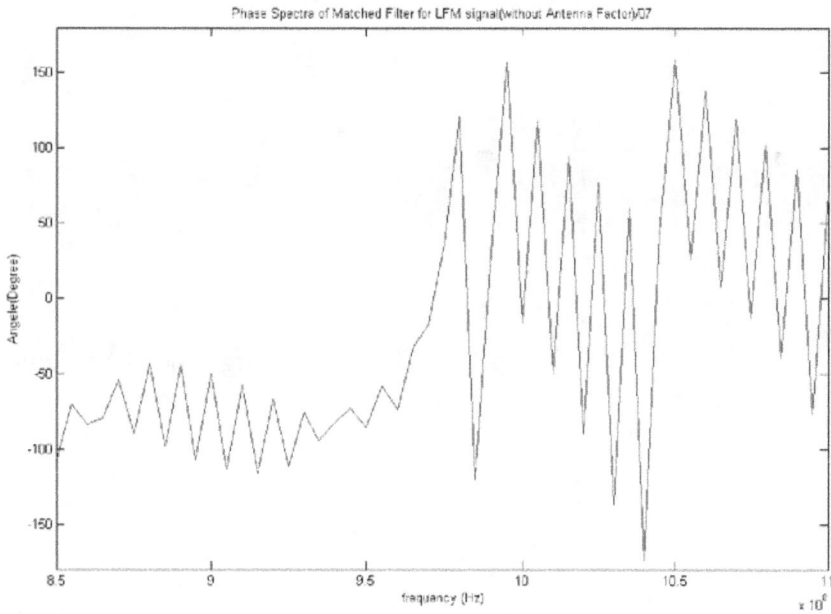

Fig.7 Phase Spectra of Desired Matched Filter (without ATF)

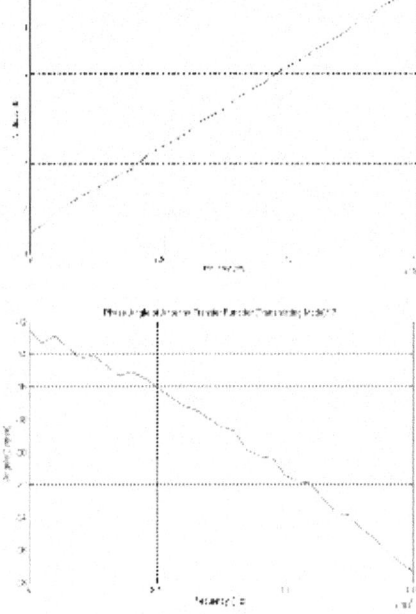

Fig 8 ATF in Transmitting Mode (Amplitude)

Fig 9 ATF in Transmitting Mode (Phase)

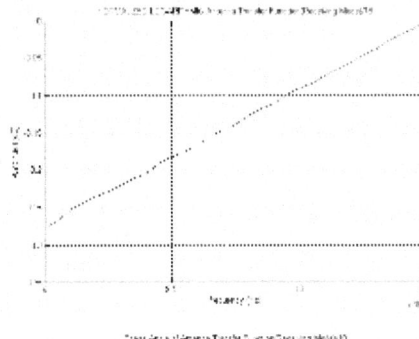

Fig 10 ATF in Receiving Mode (Amplitude)

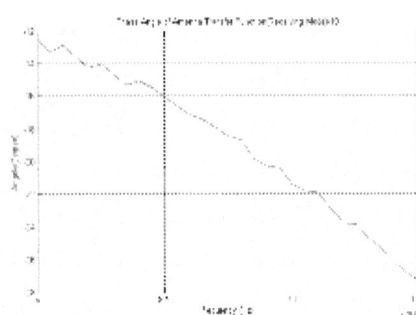

Fig 11 ATF in Receiving Mode (Phase)

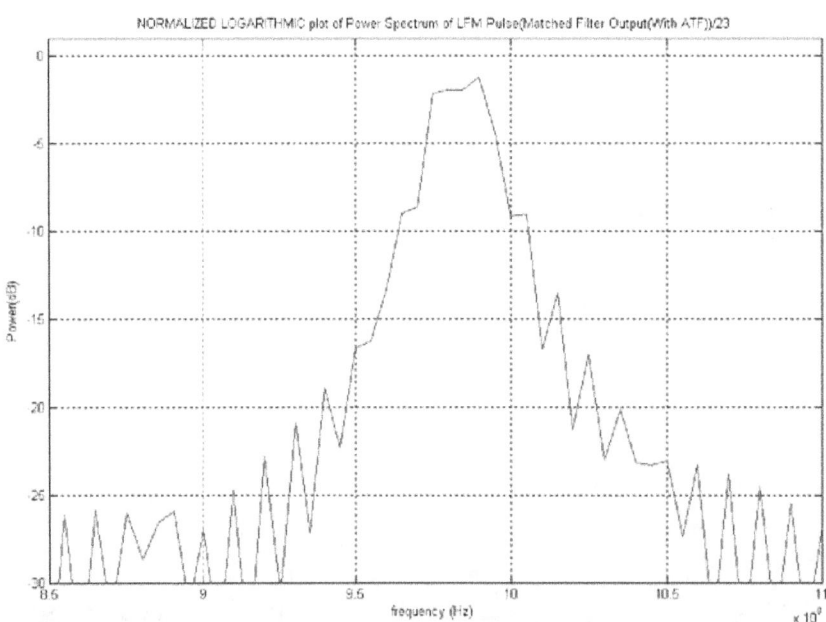

Fig.12 Matched Filter Output Power Spectra with ATF (Amplitude)

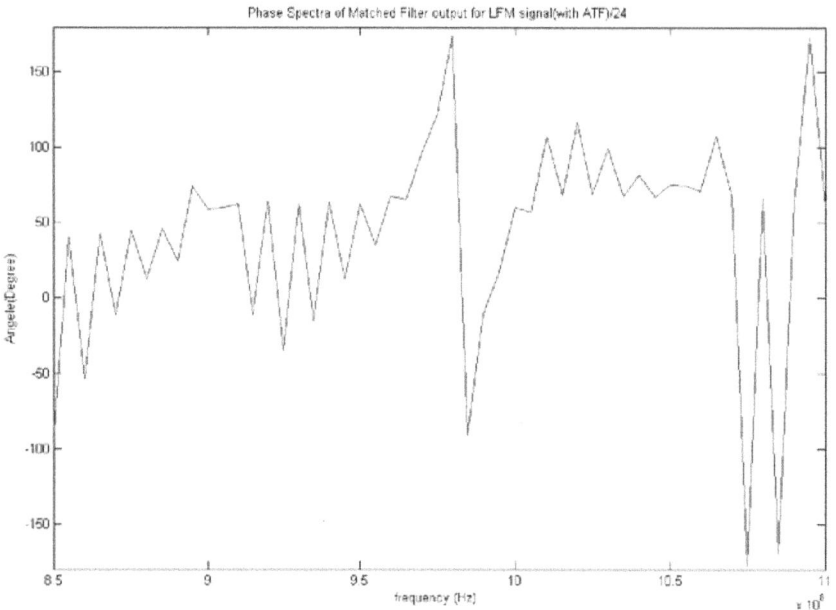

Fig.13 Matched Filter Output Power Spectra with ATF (Phase)

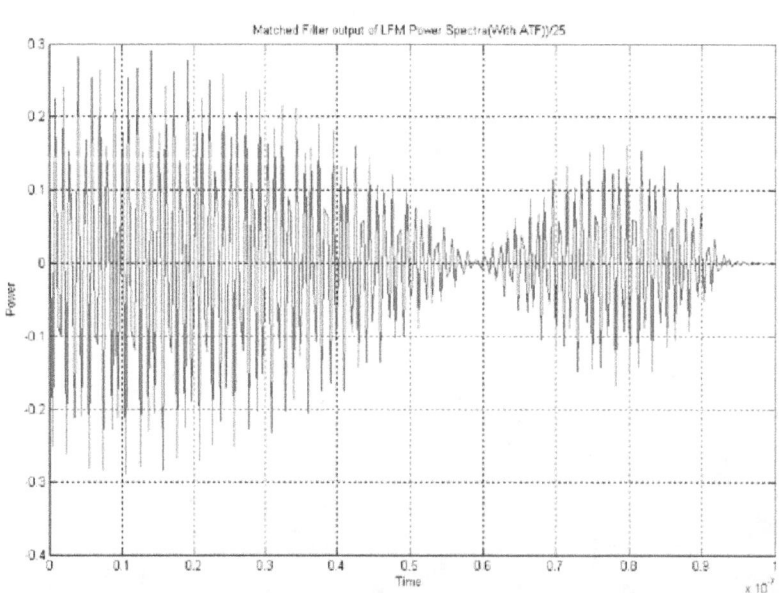

Fig.14 Time domain Matched Filter output showing no peak at t=0 with Antenna Factor
Estimation

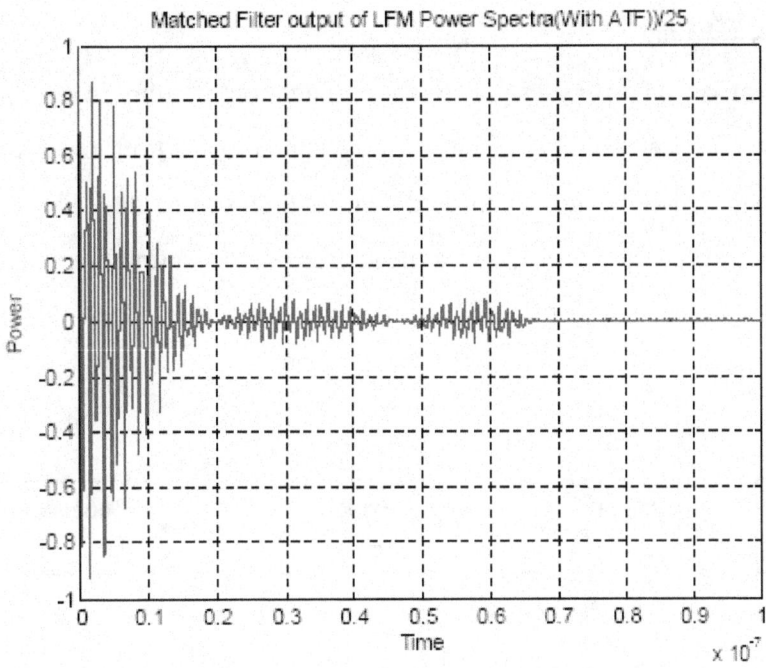

Conclusion: -

Initially we have designed a matched filter for a receiver to receive a transmitted LFM pulse without considering the Antenna Transfer Function into account. Although Simulation shows that we can correctly received the LFM pulse by using a matched filter in receiver only compensating the Channel Transfer Function, Practical results differs due to Antenna Transfer Function (ATF) and channel nonlinearity. We have consider a LTI channel for this project work and redesigned the matched filter considering ATF, which shows that, then we can receive the transmitted LFM pulse correctly. Practical results also agree with this some with small error due to channel non-linearity. For counter non-liner Channel behavior one has to consider the design of an adaptive equalizer

What is an Equalizer?

The basic problem of receiver design in communication system is the presence channel distortion, which is not known at a priori. The channel distortion results in inter symbol interference (ISI), which, if left uncompensated, causes high error rates. The solution to the ISI problem is to design a receiver that employs a means for compensating or reducing the ISI in the received signal. The compensator for the ISI is called an *equalizer*.

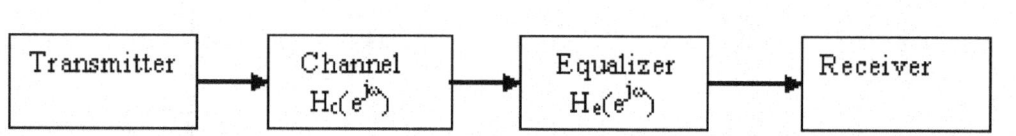

Basic communication system with equalizer

From the definition of the equalizer we have:

$$H_e(e^{j\omega}) = [\, H_c(e^{j\omega}) \,]^{-1}$$

Adaptive equalizer for Time Varying Channel equalization

In most of the communication system that employ equalizers, the channel characteristics are unknown a priori and, in many cases, the channel response is time-variant. In such a case the equalizers are designed to be adjustable to the channel response and, for time –variant channels, to be adaptive to the time variations in the channel response.

The equalizers based on algorithms for automatically adjusting the equalizer coefficients to optimize a specified performance index and to adaptively compensate for time variations in the channel characteristics are called *adaptive equalizers*.

The algorithms on which the adaptive filters are based are recursive in nature which makes it possible for the filter to perform satisfactorily in an environment where complete knowledge of the relevant signal characteristics is not available the algorithm starts from some predetermined set of *initial conditions*, representing whatever we know about the environment. In stationary conditions it is found that after successive iterations of the algorithm it *converges* to optimum Wiener solution in some statistical sense. In a non stationary environment, the algorithm offers a *tracking* capability, in that it can track time variation in the statistics of input data.

A wide variety of recursive algorithms have been developed in the literature for the operation of linear adaptive filters. In the final analysis, the choice of one algorithm over another is determined by one or more of the following factors:-

- *Rate of convergence:* this is defined as the number of iterations required for the algorithm to stationary inputs, to converge to the optimum Wiener solution in the mean square error sense.

- *Misadjustment:* this parameter provides a quantitative measure of the amount by which the final value of the mean-square error, averaged over an ensemble of adaptive filters, deviates from the minimum mean-square error produced by the Weiner filter.

- *Tracking:* the tracking performance of the algorithm, however, is influenced by two contradictory features: (1) rate of convergence, and (2) steady-state fluctuation cue to algorithm noise.

- *Robustness:* for an adaptive filter to be robust, small disturbances can result in small estimation error. The disturbances can arise from a variety of factors, internal and external of filter.

- *Computational requirement:* here the issue of concern include: (a) the number of operations required making one complete iteration of the algorithm, (b) the size of memory locations required to store the data and the program, and (c) the investment required to program the algorithm on a computer.

- *Structure:* this refers to information flow in the algorithm, determining the manner in which it is implemented in hardware form.

- *Numerical properties:* when an algorithm is implemented numerically, inaccuracies are produced due to quantization errors. In particular there are two basics issues of concern (1) numerical stability which is an inherent characteristic of an adaptive filtering algorithm, and (2) numerical accuracy which is determined by the number of bits used in numerical representation of data samples and filter coefficients. An adaptive filtering algorithm is sad to be numerically robust when it is insensitive to variation in the word length used in its digital implementation.

Linear Filter Structures

The operation of adaptive filtering algorithm involves two basic process: (1) a filtering process designed to produce an output in response to a sequence of input data, and (2) an adaptive process, the purpose of which is to provide a mechanism for adaptive control of an adjustable set of parameters used in filtering process. These processes work interactively with each other. Naturally, the choice of structure for the filtering process has a profound effect on the operation of algorithm as a whole.

The impulse response of linear filter determines the filter's memory. On this basis, we may classify linear filters into *finite duration impulse response* (FIR) and *infinite duration impulse response* (IIR) filters, which are respectively characterized by finite memory and infinite memory.

Linear Filters with Finite Memory

Three types of filter structures are themselves in the context of an adaptive filter with finite memory. There structure contains *feedforward* paths only.

1. Transversal filter

2. Lattice predictor

3. Systolic array

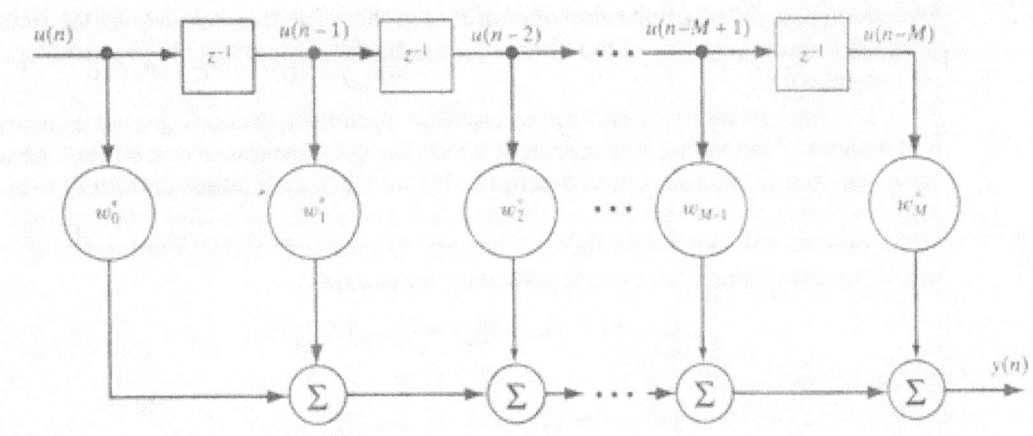

Linear Transversal Filter

Linear Filters with Infinite Memory

The feature that distinguishes an IIR filter from an FIR filter is the inclusion of *feedback* path. In fact it is the inclusion of this feedback path that makes the duration of impulse response of IIR filter infinitely long. Furthermore, the inclusion of a feedback path introduces a new problem: potential instability. By contrast, an FIR filter is inherently stable.

Approaches To The Development Of Linear Adaptive Filter

There is no unique solution to linear adaptive filter problem. Rather, we have a 'kit of tools' represented by variety of recursive algorithms. Each of which offers desirable features of its own.

Basically, we may identify two distinct approaches for driving recursive algorithms for the operation of linear adaptive filters.

Stochastic Gradient Approach

The stochastic gradient approach uses tapped-delay line, or transversal filter, as the structural basis for implementing the linear adaptive filter. For the case of stationary inputs, the *cost function*, also referred to as index of performance, is defined as the *mean square error*. This cost function is precisely a second order function of the tap weights in the transversal filter. The dependence of the mean square error on the unknown tap weights may be viewed to be in the form of multi dimensional *paraboloid* with uniquely defined bottom, or *minimum- point*. We refer to this paraboloid as error performance surface; the tap weights corresponding to the minimum point of the surface define the optimum Weiner solution.

To develop a recursive algorithm for updating the tap weights of the adaptive transversal filter, we proceed in two stages. First we use an iterative procedure to solve the *Weiner –Hoff equations*; the iterative procedure is based in the method of steepest decent, which is well known technique in optimizing theory. This method requires the use of a *gradient vector*. The value of which depends on two parameters; the *correlation matrix* of the tap inputs of the transversal filter and the *cross-correlation vector* between the desired response and the same tap input. Next, we use the instantaneous values of these correlations, so as to derive an *estimate* for the gradient vector, making it assume a *stochastic* character in general. The resulting algorithm is widely known as *least-mea-square* (LMS) *algorithm*, the essence of which for the case of traversal filters operating on real valued data, may be described as

Updated value of tap-weight vector	=	old value of tap-weight vector	+	learning- rate parameter	x	tap- input vector	x	error signal

where the error signal is defined as the difference between some desired response and the actual response of the transversal filter produced by tap input vector.

The LMS algorithm is simple yet capable of achieving satisfactory performance under right conditions. Its major limitations are a relatively slow rate of convergence and sensitivity to variation in the *condition number* of the correlation matrix if the tap inputs; the condition number of a Hermitian matrix is defined as the ratio of its largest eigenvalue to its smallest eigenvalue. Nevertheless the LMS algorithm is highly popular and widely used in a variety of applications.

In a non-stationary environment, the orientation of the error performance surface varies continuously with time. In this case, the LMS algorithm has a added task of continually tracking the bottom of the error-performance surface. Indeed, tracking will occur, provided the input data vary slowly compared with the *learning rate* the LMS algorithm.

The stochastic gradient approach may also be pursued in the context of a lattice structure. The resulting filtering algorithm is called the *gradient adaptive lattice* (GAL) *algorithm*. In their own individual ways, the LMS and GAL algorithms are just two members of the *stochastic gradient family* of linear adaptive filters, although it must be said that LMS algorithm is by far the most popular member of this family.

Least-Square Estimation

The second approach to the development of linear adaptive filtering algorithms is based on the *method of least squares*. According to this method, we minimize a cost function or an index of performance which is defined as the *sum of weighted error squares*, where the *error* or *residuals* itself defined as the difference between some desired response and the actual filter output. The method of least square can be formulated with *block estimation* or *recursive estimation* in mind. In block estimation, the input data stream is arranged in the form of blocks of equal length, and the filtering of input data proceeds on a block by block basis. In recursive estimation, on the other hand, the estimates of interest are *updated* on a sample by sample basis. Ordinarily, a recursive estimator requires less memory than a block estimator, which is the reason for which wider use of the recursive estimator in practice.

The *recursive least square family* of linear filtering algorithms may be divided into three different categories, depending on the approach taken:

1. The Standard RLS Algorithm
2. Square Root RLS Algorithm
3. Fast RLS Algorithm

Ending the introduction of adaptive filers something has to be said about the tracking behavior of the algorithms. In this context, we note that stochastic gradient algorithms such as the LMS algorithm are *model independent*; generally speaking we would expect them to exhibit good tracking behavior, which indeed they do. In contrast, RLS algorithms are *model dependent*; this, in turn, means that their tracking behavior may be inferior to that of a member of the stochastic gradient family, unless care is taken is taken to minimize the mismatch between the mathematical model on which they are based and the underlying physical process responsible for generating the input data.

Least-Mean-Square Adaptation Algorithm

Consider a transversal filter with tap inputs $u(n)$, $u(n-1)$, ..., $u(n-M+1)$ and a corresponding set of tap weights $w_0(n)$, $w_1(n)$, ..., $w_{M-1}(n)$. The tap inputs represent samples drawn from a wide-sense stationary stochastic process of zero mean and correlation matrix **R**. in addition to these inputs, the filter is supplied with a desired response $d(n)$ that provides a frame of reference for optimum filtering action.

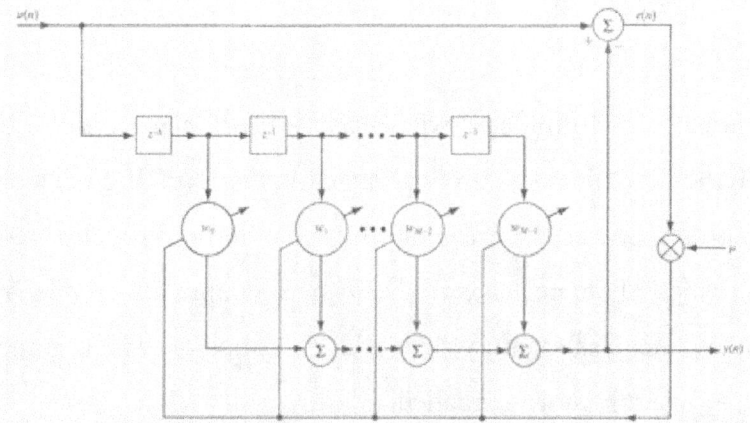

Linear Transversal Filter in Training Mode

The vector of tap inputs at time n is defined by $\mathbf{u}(n)$, and the corresponding estimate of the desired response at the filter output is denoted by $\mathbf{d}(n)$. By comparing this estimate the desired response $d(n)$, we produce an estimation error denoted by $e(n)$. We may write

$$e(n) = d(n) - \mathbf{d}(n)$$

$$= d(n) - \mathbf{w}H(n)\mathbf{u}(n)$$

where the term $\mathbf{w}H(n)\mathbf{u}(n)$ is the inner product of the tap-weight vector $\mathbf{w}(n)$ and the tap-input vector $\mathbf{u}(n)$. The expanded form of tap weight vector is defined by

$$\mathbf{w}(n) = [w_0(n), w_1(n), \ldots, w_{M-1}(n)]^T \; ,$$

and the tap-input vector is described by

$$\mathbf{u}(n) = [u(n), u(n-1), \ldots, u(n-M+1)]^T \; .$$

If the tap-input vector $\mathbf{u}(n)$ and the desired response $d(n)$ are jointly stationary, then the mean square error or cost function $J(\mathbf{w}(n))$, or simply $J(n)$, at time n is a quadratic function of the tap-weight vector, so we may write:

$$J(n) = \sigma_d^2 - \mathbf{w}H(n)\mathbf{p} - \mathbf{p}^H\mathbf{w}(n) + \mathbf{w}H(n)\mathbf{R}\mathbf{w}(n),$$

where σ_d^2 = variance of the desired response $d(n)$,

$\quad\quad \mathbf{p}$ = cross-correlation vector between the tap-input vector $\mathbf{u}(n)$ and the desired

$\quad\quad\quad\quad$ response $d(n)$, and

$\quad\quad \mathbf{R}$ = correlation matrix of the tap-input vector $\mathbf{u}(n)$.

$$\nabla J(n) = -2\mathbf{p} + 2\mathbf{R}\,\mathbf{w}(n)$$

The simplest choice of estimators is to use instantaneous estimates for $\mathbf{R}$ and $\mathbf{p}$ that are based on sample values of the tap-input vector and desired response, defined respectively by

$$\mathbf{R}(n) = \mathbf{u}(n)\mathbf{u}^H(n)$$

and $\quad\quad\quad\quad\quad\quad \mathbf{p}(n) = \mathbf{u}(n)\, d*(n).$

Correspondingly, the instantaneous estimate of the gradient vector is

$$\nabla J(\text{n}) = -2\mathbf{u}(\text{n})d^*(\text{n}) + 2\mathbf{u}(\text{n})\mathbf{u}^{\text{H}}(\text{n})\mathbf{w}(\text{n}).$$

Generally speaking, this estimate is biased, because the tap-weight estimate vector $\mathbf{w}(\text{n})$ is a random vector that depends on the tap-input vector $\mathbf{u}(\text{n})$. Note that the estimate $\nabla J(\text{n})$ may also be viewed as the gradient operator ∇ applied to the instantaneous squared $|e(\text{n})|^2$.

Using the estimate of the gradient vector we get a recursive relation for updating the tap-weight vector:

$$\mathbf{w}(\text{n}+1) = \mathbf{w}(\text{n}) + \mu\mathbf{u}(\text{n})[d^*(\text{n}) - \mathbf{u}^{\text{H}}(\text{n})\mathbf{w}(\text{n})]$$

where μ is a positive constant called the step-size parameter.

From this we have,

1. Filter output:
$$y(\text{n}) = \mathbf{u}^{\text{H}}(\text{n})\mathbf{w}(\text{n}).$$

2. Estimation error or error signal:
$$e(\text{n}) = d(\text{n}) - y(\text{n}).$$

3. Tap-weight adaptation:

$$\mathbf{w}(\text{n}+1) = \mathbf{w}(\text{n}) + \mu\mathbf{u}(\text{n})e^*(\text{n}).$$

At each iteration or time update, this algorithm requires knowledge of the most recent values: $\mathbf{u}(\text{n})$, $d(\text{n})$, and $\mathbf{w}(\text{n})$.

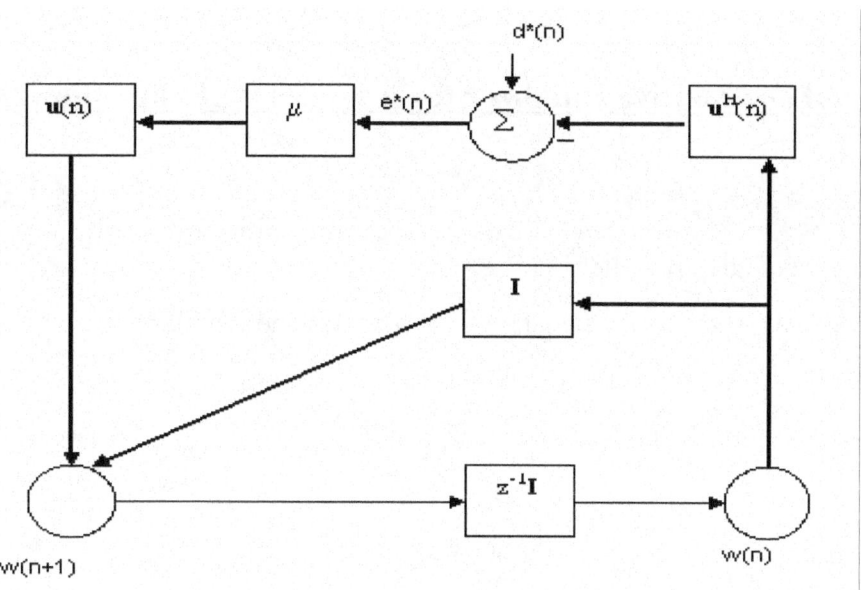

Signal flow graph representation of the LMS algorithm

Note: Typically, adaptive equalizers used in digital communication require an initial training period, during which a known data sequence is transmitted. A replica of this sequence is made available at the receiver in proper synchronization with the transmitter, thereby making it possible for adjustment to be made to the equalizer coefficients in accordance with adaptive filtering algorithm employed in the equalizer design. When the training is completed, the equalizer is switched to its decision directed mode, and normal data transmission may then commence.

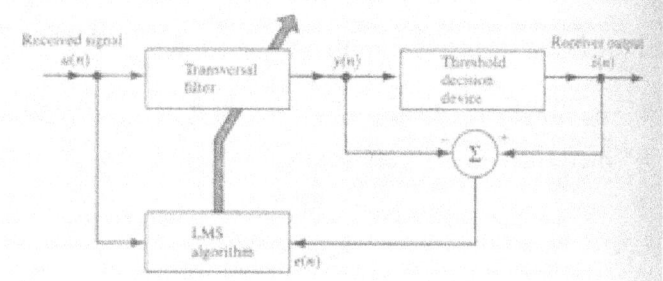

Linear Transversal Filter in Decision Directed mode

Case 1: Design of an adaptive equalizer for a system with the following specifications

- The degree of adaptive filter is 11.
- The impulse response of the channel is a raised cosine function;

$$h_n = (\tfrac{1}{2})[1+\cos(2\pi(n-2)/w)], \quad n=1,2,3$$

$$0 \qquad\qquad\qquad\qquad \text{otherwise}$$

- The signal applied to this channel consists of a Bernoulli sequence taking values 1 or 0 with probability 0.5 each.
- The additive noise is Gaussian noise with zero mean, and variance $\sigma^2 = 0.001$.
- The step size parameter of the adaptation algorithm (the filter) is $\mu_1 = 0.08$ for the training mode and $\mu_2 = 0.0003$ for decision mode directed mode.
- The adaptive filter is working on training mode for the first 1000 samples.
- The adaptive filter is working in decision directed mode for the samples between 1000 samples and 5000 samples.
- The initial filter coefficients are equated zero. At each iteration these coefficients are modified and at the beginning of the decision directed mode the filter coefficients of the last iteration of training mode are taken as the initial coefficients.
- The equalized signal is passed through a slicer; the slicer is actually a quantizer. The rule of quantization is that it quantizes the signal to 1 when the signal is greater than 0.5 and quantizes to 0 when the signal is less than 0.5.

Output:

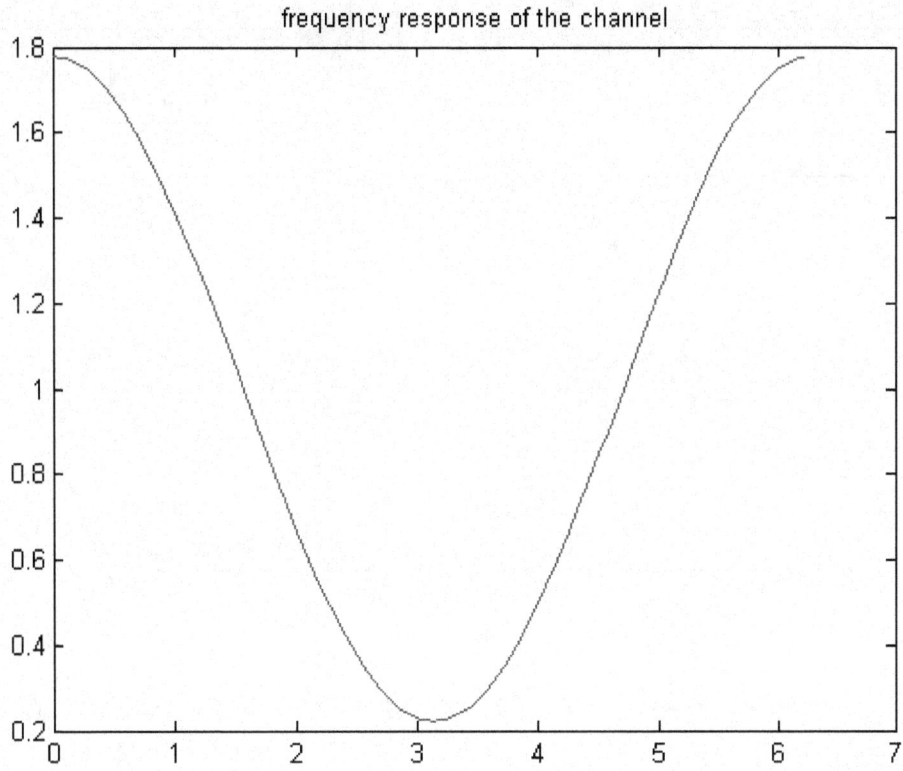

Frequency response of the unknown channel taken here to be a raised cosine.

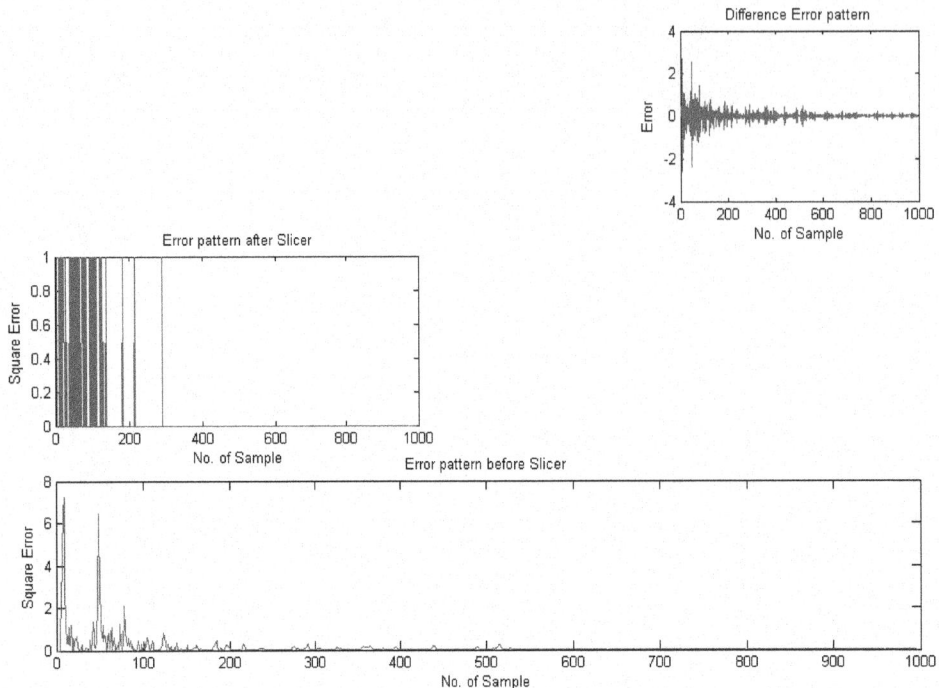

Here the error at different stages of the system after *training mode* is displayed. They are:

(a) Error pattern before the slicer.

(b) Error pattern after the slicer.

(c) Difference error pattern.

It is observed that error pattern of the system becomes zero after about 250 iterations i.e. after 250 iterations the equalizer has adapted to the channel.

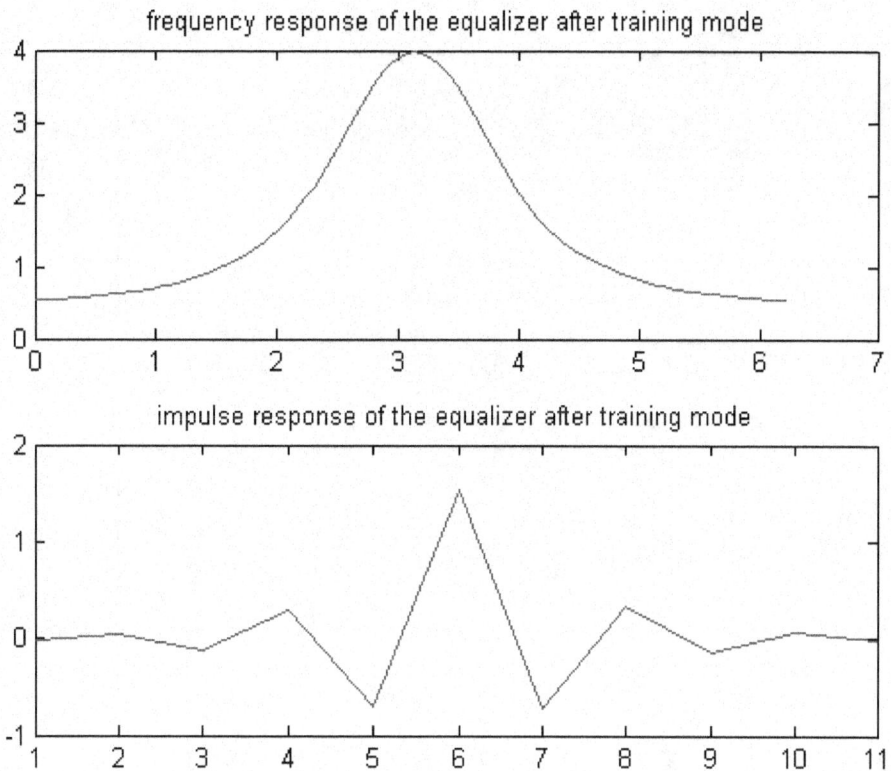

From the frequency response of the equalizer, after the *training mode*, it can be observed that our adaptive equalizer has very well adapted to the channel and the satisfying the basic equalizer condition mentioned in the beginning:

$$H_e(e^{j\omega}) = [\, H_c(e^{j\omega})\,]^{-1}$$

Where, $H_e(e^{j\omega})$ is the frequency response of the equalizer, and

$H_c(e^{j\omega})$ is the frequency response of the channel.

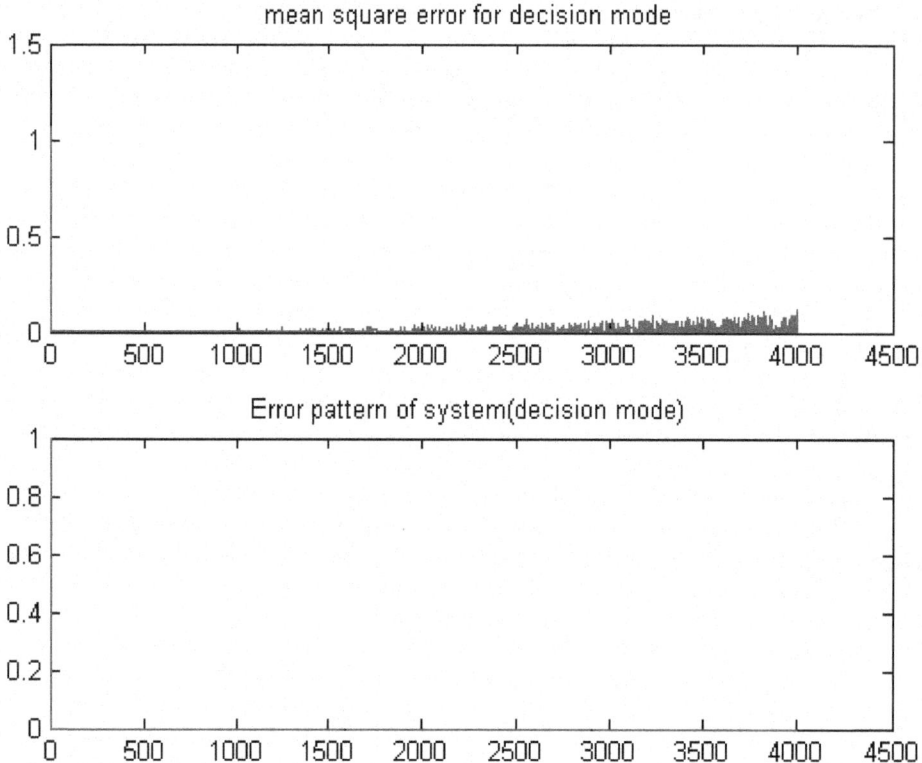

From the error patterns of the system, in *decision mode* it is seem the system error zero i.e. our equalizer is adapted to the channel and is nullifying the effect of the channel. However, the mean square error pattern is diverging but is well within limits and is causing no harm to the system output due to which we have zero system error. But if working in this mode is allowed for a longer time mean square error would diverge enough to introduce system error. So the filter has to be made to operate in the training mode again to match it to channel.

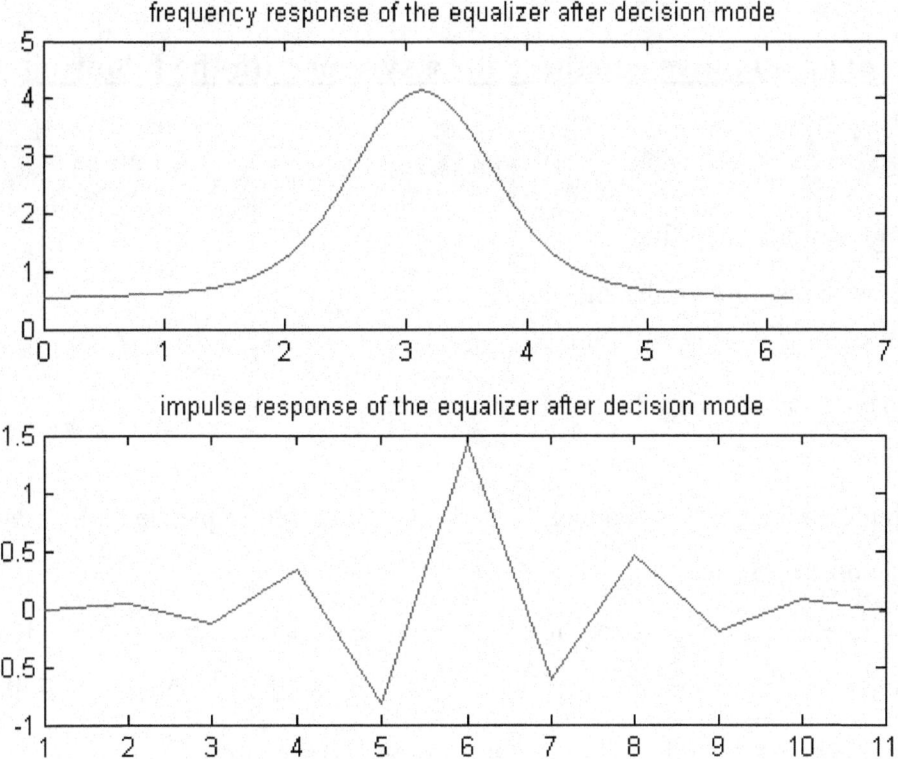

The frequency response of the equalizer, after the *decision mode*, shows that it is still matched to the channel agreeing to the fact that we are not getting any system error in the decision mode.

Case 2: Design of an adaptive equalizer for a system with the following specifications

- The degree of adaptive filter is 11.
- The impulse response of the channel is

$$h_n = [\ 0.2, -0.15, 1.0, 0.21, 0.03\ ].$$

  this channel has a null in the mid frequency region.

- The signal applied to this channel consists of a Bernoulli sequence taking values 1 or 0 with probability 0.5 each.
- The additive noise is Gaussian noise with zero mean, and variance $\sigma^2 = 0.001$.
- The step size parameter of the adaptation algorithm (the filter) is $\mu_1 = 0.08$ for the training mode and $\mu_2 = 0.0003$ for decision mode directed mode.
- The adaptive filter is working on training mode for the first 1000 samples.
- The adaptive filter is working in decision directed mode for the samples between 1000 samples and 5000 samples.
- The initial filter coefficients are equated zero. At each iteration these coefficients are modified and at the beginning of the decision directed mode the filter coefficients of the last iteration of training mode are taken as the initial coefficients.
- The equalized signal is passed through a slicer; the slicer is actually a quantizer. The rule of quantization is that it quantizes the signal to 1 when the signal is greater than 0.5 and quantizes to 0 when the signal is less than 0.5.

output:

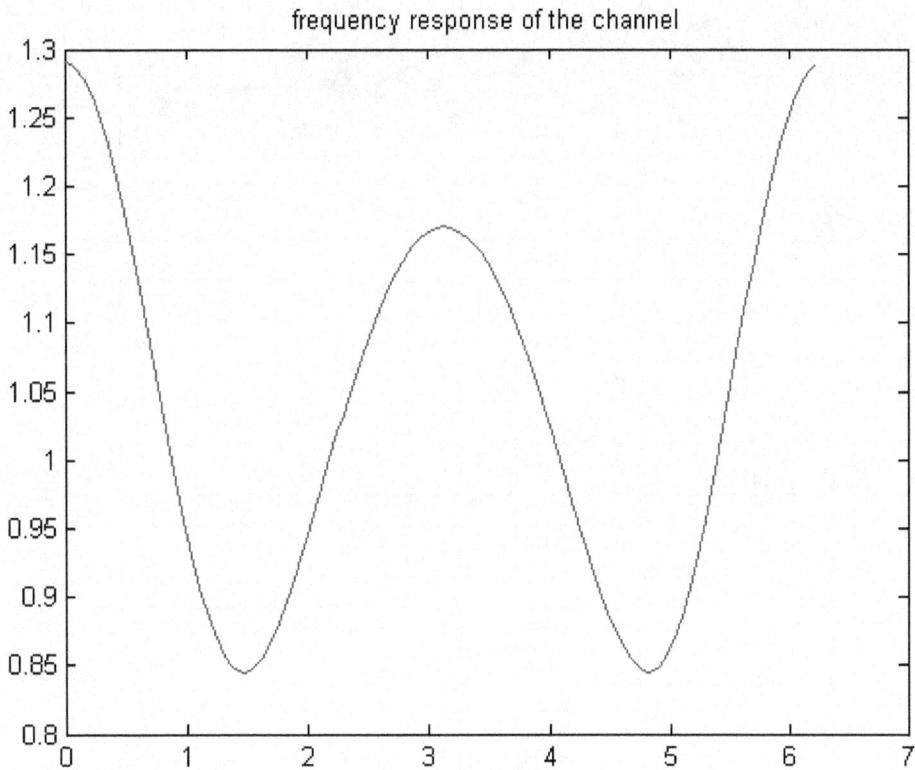

Frequency response of the unknown channel to which our equalizer has to adapt.

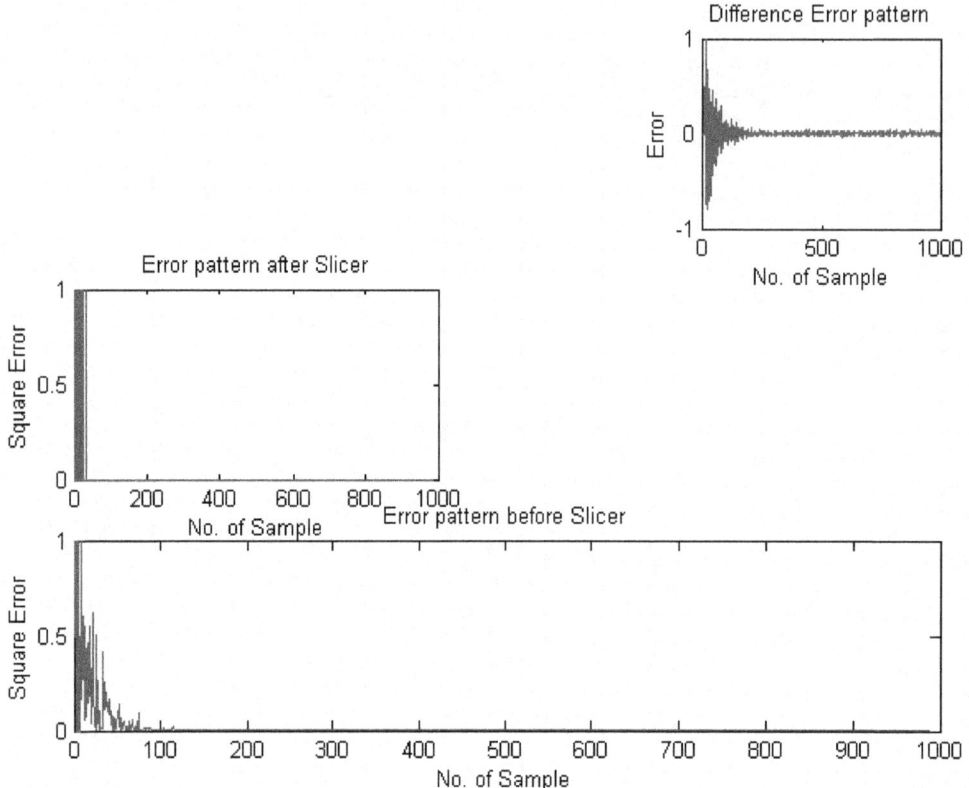

As was seen in the first case, in the second case as well the equalizer has very well adapted to the channel during the training mode. The error is converging as can be seen that error pattern after the slicer i.e. system error has gone to zero after 50 samples.

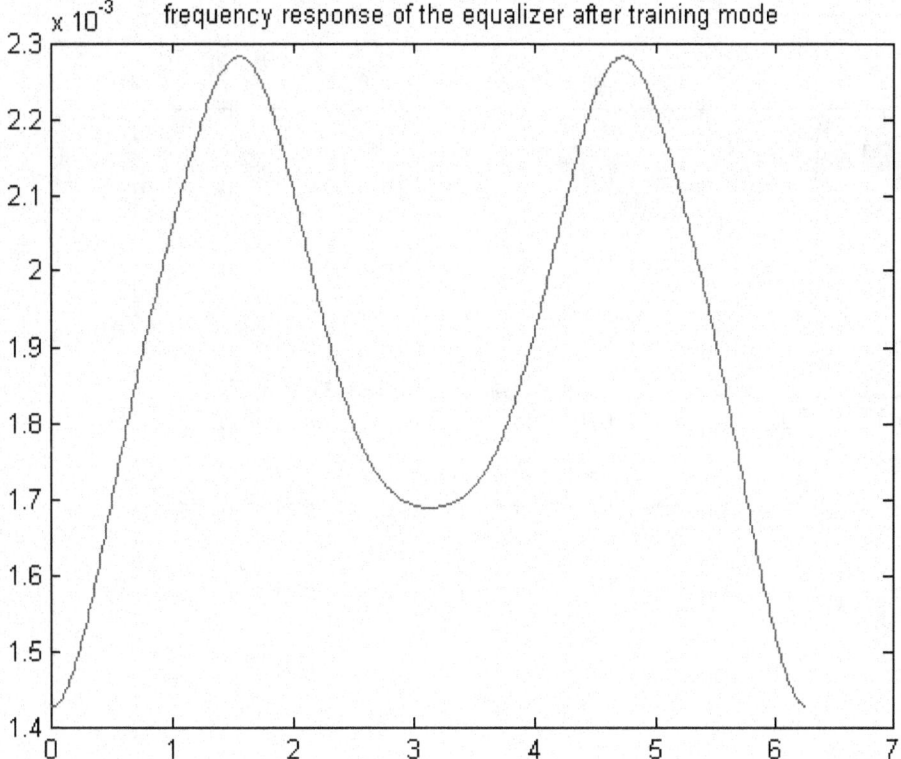

The frequency response of the equalizer also shows that it has adapted to the channel and as was in the first case it now satisfies the basic equalizer condition.

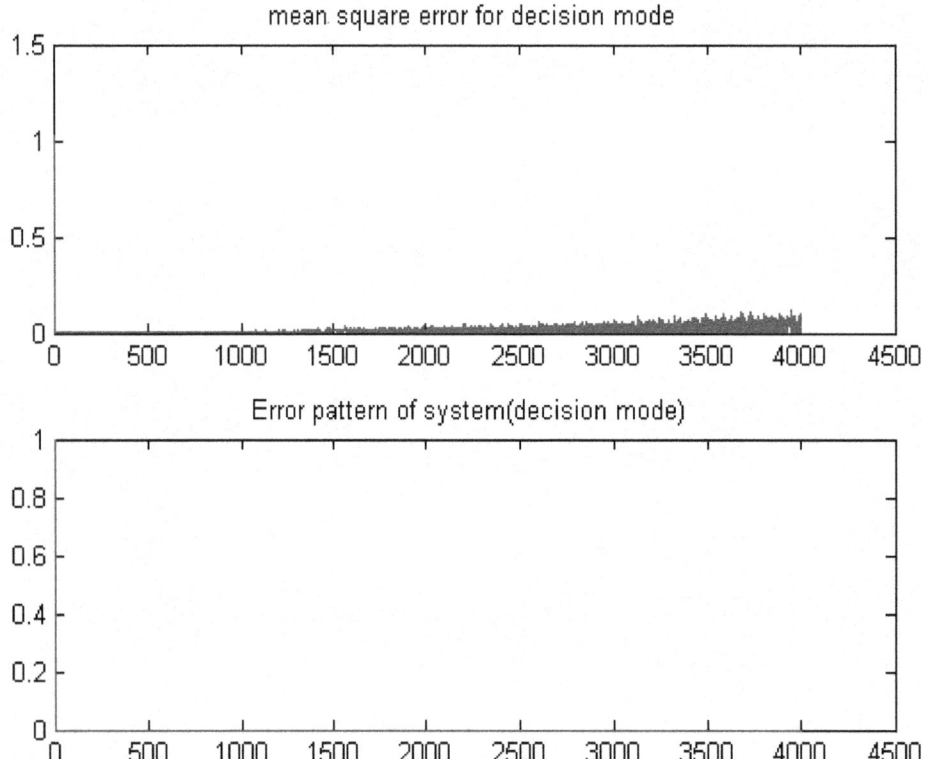

Similar to the case 1 we observe that the error is diverging but is not enough to cause any system error but the system has to be switched to training mode before this error starts affecting the output.

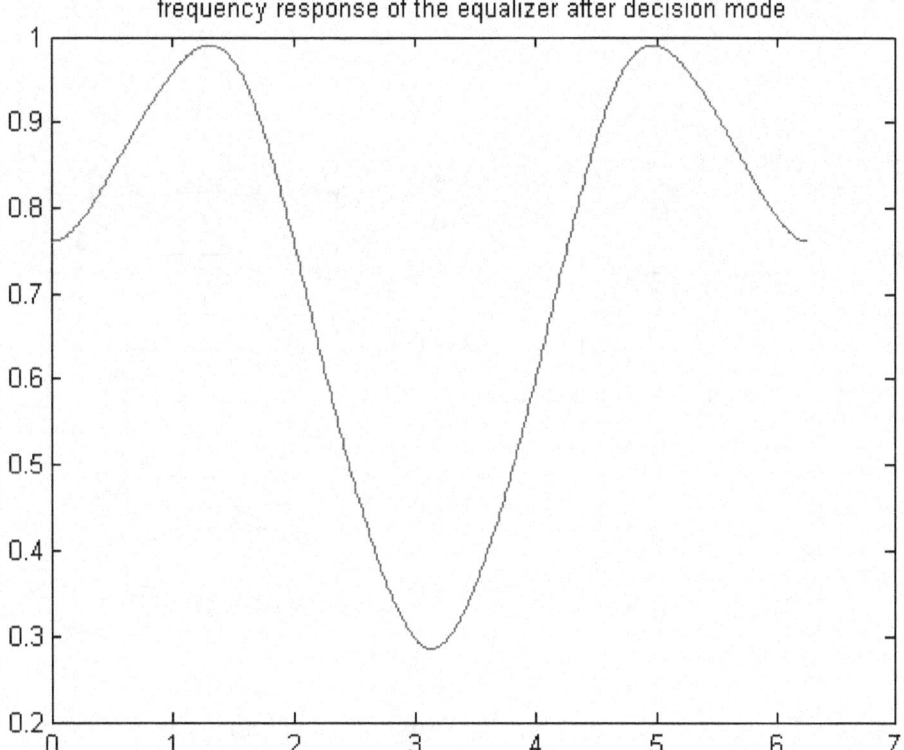

One important difference at this point between the first case and this case is that in the first case at his point the equalizer was matched as good as it was at the end of training mode but here the equalizer has diverged from what it was at the end of training mode.

The reason for this the spectral null is the channel. Due to this the equalizer response diverges from the training mode response much faster than in the previous case.

Adaptive Decision Feedback Equalizer (DFE)

- A decision feedback equalizer (DFE) is a non-linear equalizer that uses detector decisions to eliminate the ISI on pulses that are currently being demodulated.

- The basic idea of DFE is that if values of the symbols previously detected are known (past decisions are assumed to be correct), then the ISI by these symbols can be cancelled out of the forward filter by subtracting past symbol values with appropriate weighting. The figure below shows a general structure of adaptive feedback equalizer.

59

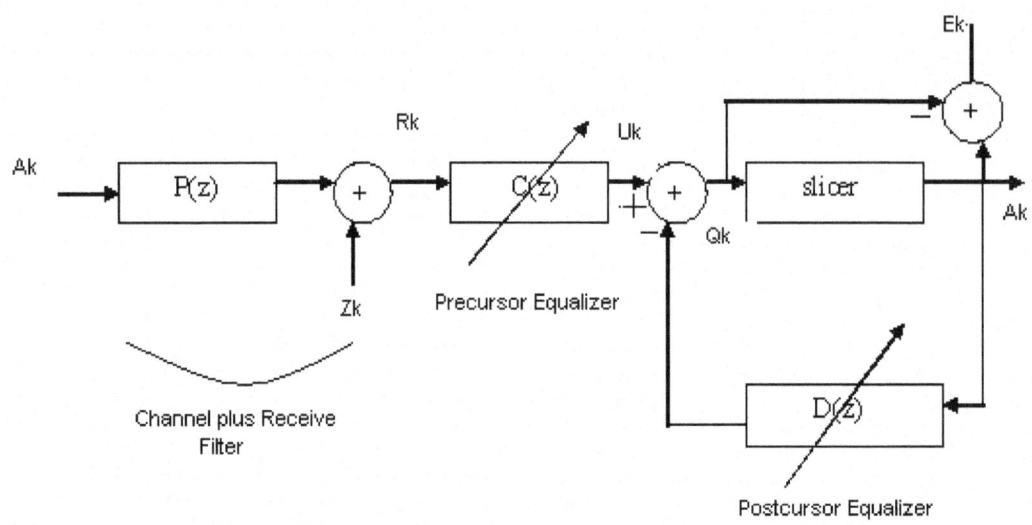

Block diagram of the adaptive decision feedback equalizer

- If we look at the figure, we see that the estimated signal sequence becomes,

$$Q_k = \sum_{i=-(N-1)}^{0} c_i R_{k-I} - \sum_{i=1}^{M} d_i A_{k-i}$$

- $\{c_i\}$'s coefficients of the precursor equalizer, $\{d_i\}$'s coefficients of the post cursor equalizer. N is the number of precursor equalizer coefficients and M is the number of post cursor equalizer coefficients.

- Adaptive DFE algorithm is similar to stochastic gradient algorithm, with the important difference that the input to the causal portion of the filter is the decisions rather than the output of the precursor equalizer filter.

- This difference will obviously change the desired tap coefficients as well as reduce the noise due to equalization.

- The derivation of a stochastic gradient algorithm for the DFE is a simple extension of the stochastic gradient algorithm for the linear case. First, we define an augmented vector of N+M coefficients,

$$V^T = [c_{-(N-1)} \ldots\ldots c_0 \ -d_1 \ldots\ldots -d_M]$$

- And augmented input signal vector

$$W_k^T = [R_{k+(N-1)} \ldots\ldots R_k \ A_{k-1} \ldots\ldots A_{k-M}]$$

- DFE slicer error can be expressed as,

$$E_k = a_k - V_k^T W_k^T$$

- The adaptation algorithm becomes,

$$V_{k+1} = V_k + \beta E_k W_k^*$$

Case: Design of Adaptive Decision Feedback Equalizer for the system with following specifications.

- The degree of the precursor equalizer is 5, and the degree of post cursor equalizer is 3.
- The impulse response of the channel is [0.2, -0.15, 1.0, 0.21, 0.03]. The channel has a spectral null in the middle frequency region.
- The signal applied to the channel consists of a Bernoulli sequence taking he values 1 and 0 with probability 0.5 each.
- The adaptive noise is Gaussian noise with mean zero, and variance 0.001.

- The step size parameter of the precursor adaptation algorithm is β_{pre_1}=0.08 for the training mode and β_{pre_2}=0.00008 for the decision mode, and step size parameter of the post cursor adaptation algorithm is β_{post_1}=0.000001 for the training mode and β_{post_2}=0.00002 for the decision mode.
- The adaptive filter is working on training mode for the first 1000 samples.
- The adaptive filter is working on decision directed mode for the samples between 1000 and 5000.
- The initial filter coefficients are equated to zero. At each iteration these coefficients are modified and at the beginning of decision directed mode the filter coefficients of the last iteration of the training mode are taken as initial coefficients.
- The equalized signal is passed through the slicer; the slicer is actually a quantizer. The rule of the quantizer in this simulation is that it quantizes the signal greater than 0.5 to 1 and quantizes the signal to 0 when it is less than 0.5.

Output:

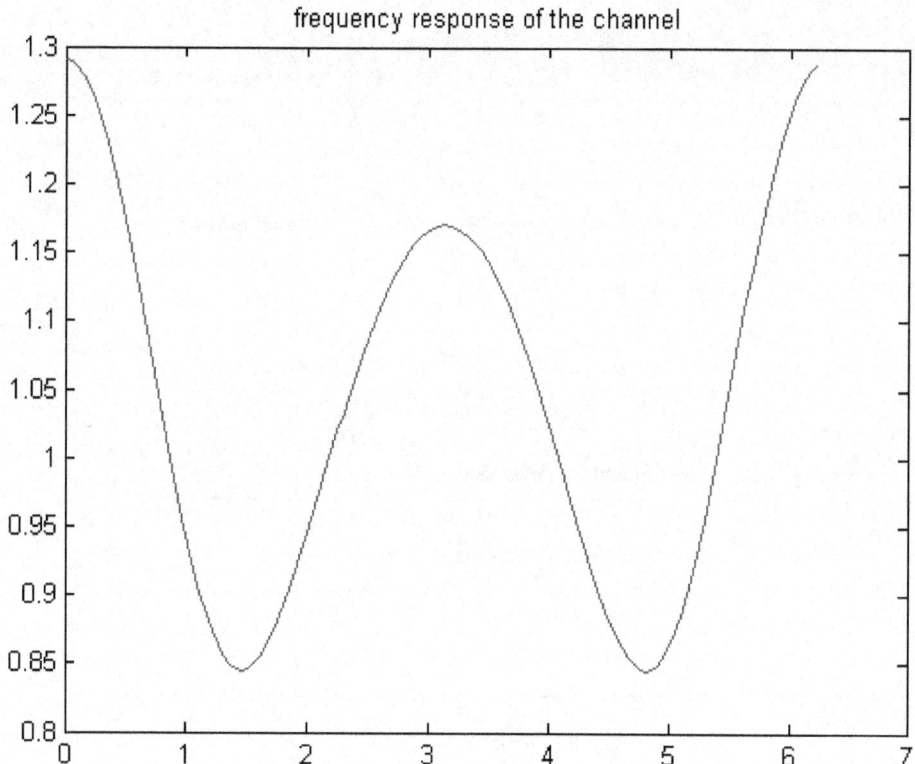

Frequency response of the unknown channel to which our equalizer has to adapt.

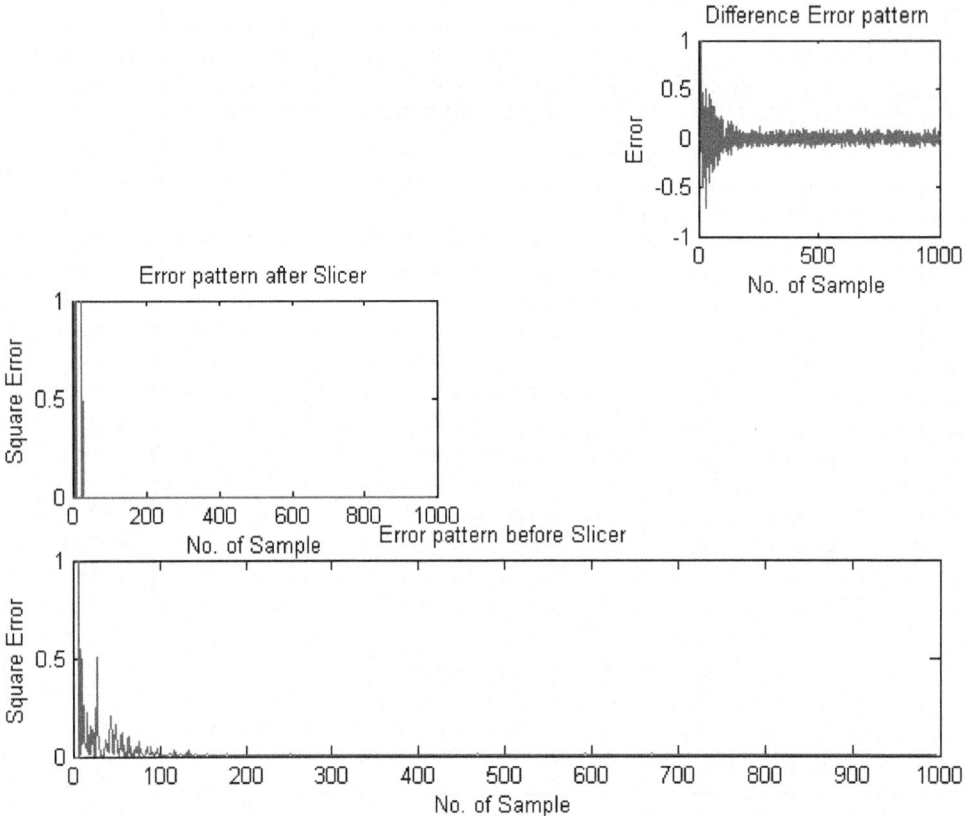

From the different error patterns it is visible that the equalizer has adapted to the channel.
The difference error pattern is converging and the error pattern after the slicer that is the error
pattern of the system has reduced to zero.

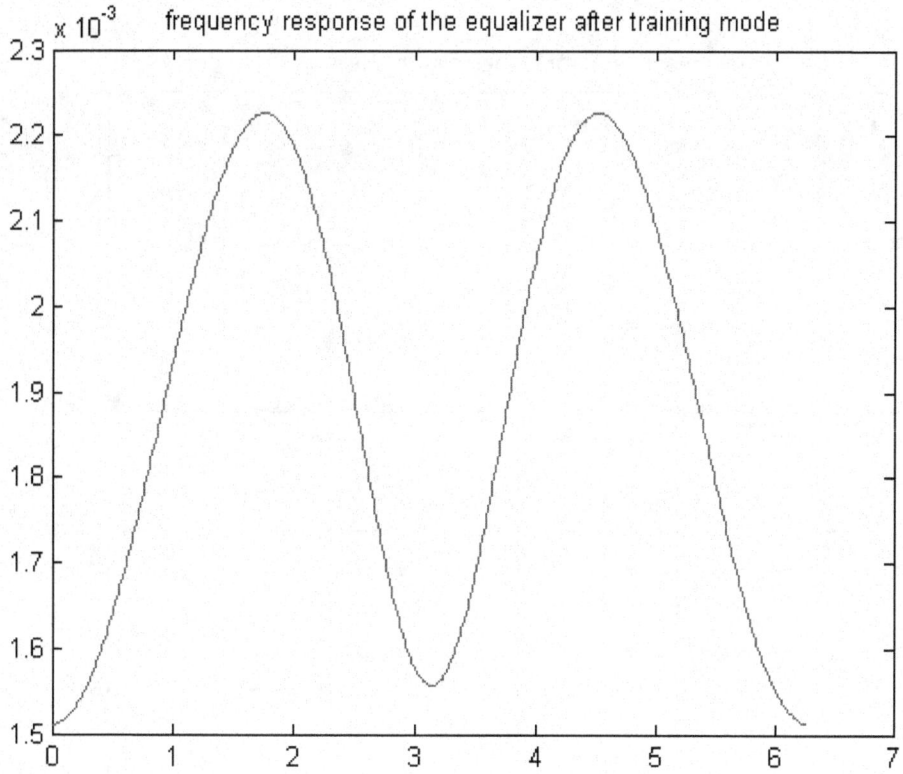

In the previous figure we saw that the error had gone to zero after some 30 iterations indicating that the equalizer has adapted to the channel. Same is visible from this figure which is the frequency response of the filter after the training mode. Comparing this figure with the frequency response of the channel it is seen that both are inverse of each other indicating that our equalizer has adapted to the channel well and is nullifying the effect of the channel which is also reflected from the error pattern after the slicer.

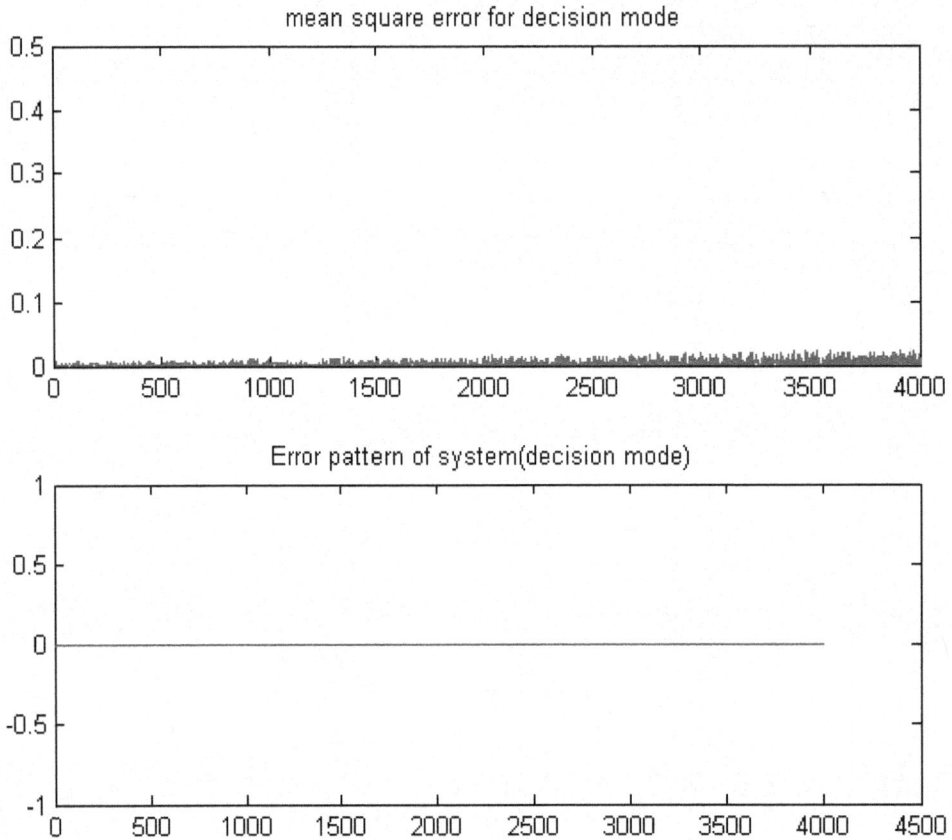

From the above figure it is observed that our equalizer which had adapted the channel in the training mode is functioning very well in the decision directed mode and there is no error in the detection of the data and we have zero system error throughout the decision directed mode. The mean square error is a bit clumsy but it has no effect on the system performance.

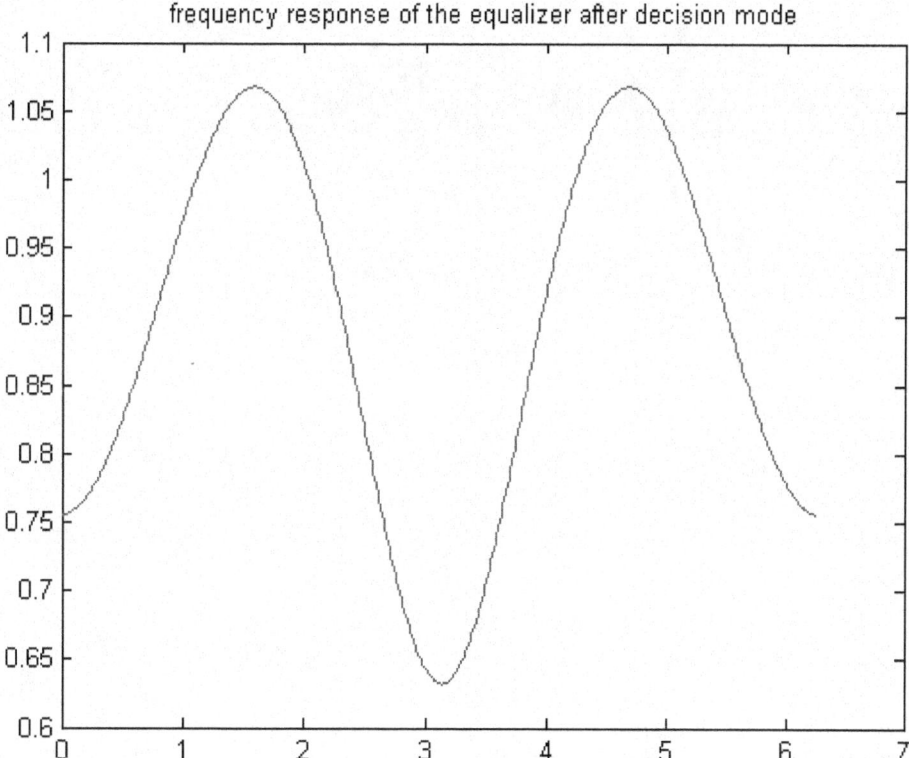

frequency response of the equalizer after decision mode

The frequency response of the equalizer after the decision directed mode shows that our filter is still very well adapted to the channel.

Here I would like to compare this result with the one of the equalizer without decision feedback discussed in case 2. There the response of the channel had diverted away from the one desired much more as compared to this decision directed equalizer case.

Also from the simulation it was observed that what was the condition of the equalizer after 5000 samples in the case without decision feedback, decision feedback equalizer has gone up to 20000 samples before degrading to that level which is surely an advantage over the previous equalizers.